Changing Gears

Changing Gears

The Strategic Implementation of Technology

James Carlopio

First published 2003 by
PALGRAVE MACMILLAN
Houndmills, Basingstoke, Hampshire RG21 6XS and
175 Fifth Avenue, New York, N.Y. 10010
Companies and representatives throughout the world

PALGRAVE MACMILLAN is the global academic imprint of the Palgrave Macmillan division of St. Martin's Press, LLC and of Palgrave Macmillan Ltd. Macmillan® is a registered trademark in the United States, United Kingdom and other countries. Palgrave is a registered trademark in the European Union and other countries.

ISBN 1–4039–0482–0

This book is printed on paper suitable for recycling and made from fully managed and sustained forest sources.

A catalogue record for this book is available from the British Library.

Library of Congress Cataloging-in-Publication Data
Carlopio, James
    Changing gears : the strategic implementation of technology / James Carlopio.
      p. cm.
    Includes bibliographical references and index.
    ISBN 1–4039–0482–0 (cloth)
      1. Technological innovations. I. Title.
    T173.8 .C365 2003
    338.9'27–dc21
                                                    2002190988

10 9 8 7 6 5 4 3 2 1
12 11 10 09 08 07 06 05 04 03

Printed and bound in Great Britain by
Creative Print & Design (Wales), Ebbw Vale

CONTENTS

PART II IMPLEMENTATION

Contents

The focus of this book is the implementation of new technology, strategy, business models and other innovations. It takes a strategic look at change management and technology implementation. It crosses the boundaries of change management, technology implementation and organisational strategy.

In order to implement new technology successfully we need to distinguish clearly between implementation and installation. Something is installed when it is technical connected or operational. You can install software on 100 personal computers, but the software is not implemented until people are using most of its functionality. If the software has been installed, but people have not been trained how to use it, it is not implemented. If the software has been installed, but reward and measurement systems still 'force' people to use a different method, it is not implemented. If the software has been installed, but the organisation structure and/or culture influence people not to use certain functions or not to use it at all, it is not implemented. It is the same with non-technical changes. Once a new organisational chart has been drawn up, we can say it has been 'technically installed', but it certainly has not been implemented. If you are conducting a business process engineering exercise, once the 'as is' and 'should be' processes have been mapped and drawn up, you can say they have been 'technically installed'. They are, however, not properly implemented until people start behaving differently and the new 'should be' process is being followed.

Most people mistakenly think of implementation as a difficult, but relatively well understood, technical roll-out or installation process. The focus of this book, on the other hand, is to try to persuade you to think of implementation as a complex social process requiring individuals and organisations to innovate and change to utilise fully the capacity available within a technology, a business model or another innovation. Technical roll-out and installation are only a small part of the implementation process and are not the main determinants of success (cf. Abrahamson, 2000; Beer and Nohria, 2000; Carlopio, 1998; Goleman, Boyatzis and McKee, 2001; Piderit, 2000).

The social side of the implementation process begins when we signal our intentions to change, and start to create sensitivity to – and a sense of urgency for – the need to change. Whether we are changing individuals, groups and/or entire organisational systems (e.g., reward and recognition or measurement systems, organisational culture or structure), awareness of the need to change is a necessary first step. In order to enable change to take place, you must provide new skills, new structures and sufficient resources to motivate people to behave in new ways. A basic premise in social psychology is that behaviour is always a function of both the person (e.g., their personality and genetic inheritance) and the situation (e.g., the social and physical environment) (Lewin, 1935).

This may be represented as follows:

$$B = f(P \times S)$$

where B = behaviour, f = 'is a function of', P = the person, their personality, X = 'interaction', and S = the social situation or the environment in which the person is behaving.

In other words, people do what they do not just because that is 'how they are'. While an individual's personality does have an effect on that person's behaviour, the situation or social environment also plays a part. Consider for a moment some people who are resisting a change you have proposed. Why are they resisting? If you believed it was their personality or 'who they were' that was the sole cause of the resistance, you might conclude that they were resisting because they were 'jerks' or 'idiots' or 'inflexible'.

All of these are personal attributions based on the idea that their behaviour is caused solely by their personality. Since that is how those people are, the only things you can do to deal with their resistance are, first, try to change some fundamental aspects of their personality; second, force them to go along with your plans; or get rid of them. I am suggesting, however, that behaviour is never solely caused by someone's personality. I am suggesting that behaviour is always a function of both the person/personality and the situation/environment. If people are resisting change, from this perspective, we must consider both who they are (e.g., inflexible or flexible) and what is going on in the social and organisational environment. It could be that they are resisting the change because they are being rewarded for behaviour that is antithetical to your change (e.g., you want them to work as part of a team, but they are measured and rewarded for individual performance), or that the change causes them to physically and/or mentally 'go well out of their way' (e.g., it is hard for them and takes more energy and effort) and they are resisting for those reasons, even though they are basically quite flexible individuals.

The situation is a powerful determinant of behaviour in organisations. Reward and recognition systems, measurement systems, organisation structure and culture are all situational or environmental factors that have profound effects on people's behaviour (cf. Bolman and Deal, 1997; Burns and Stalker, 1961; Collins and Porras, 1998; Donaldson, 1985; Schiemann and Lingle, 1997; Stajkovic and Luthans, 2001). There are so many situational factors reinforcing the status quo in organisations that making change happen requires enabling processes (i.e., facilitating structures) that are well defined, well funded and well supported in order to change both the person (e.g., how they are thinking, what they know) and the situational impacts on them (e.g., reward and recognition systems, measurement systems, organisation structure and culture, project plans, role models, communications, resources).

The Chapters in this Book

In the first part of this book (Part 1, Chapters 1–3), we begin with a look at the strategic preparation process. In Chapter 1 we explore ways to gain strategic knowledge and awareness of the external organisational environment and of our internal strengths, weaknesses and opportunities. In Chapter 2 we focus on technology analysis and selection. In Chapter 3 we look at the decision process related to technology acquisition.

In the second part of the book (Part 2, Chapters 4–9), we make the move to a more local kind of knowledge and awareness in Chapter 4. Those who will be responsible for, and affected by, the changes must be told why these decisions have been made and given appropriate opportunity for input. These decisions must be linked to strategic (as well as business and more local) issues, and eventually translated into operational plans and activities.

There are a host of issues to consider at this point in the implementation process related to planning, roles, analyses and structures. After this preparation, the persuasion process begins as we sell the idea and cost-justify it. When this is complete, and is added to adequate social-psychological facilitating structures (Chapter 5), and adequate knowledge and awareness, we approach the decision and commitment phases (Chapter 6). The changes are then installed over time (Chapter 7), systems are slowly converted, and people make thousands of incremental adjustments. Finally, the changes are fine tuned, confirmed (or rejected), and then slowly embedded in local systems (Chapter 8). At this point, the micro/local implementation process blurs into the more macro/strategic innovation process as we enter a cycle of never-ending implementation, evaluation, modification and evolution, and we begin to look beyond implementation (Chapter 9).

If changes are handled in this manner, the process is respectful of both the needs of the organisation (making the changes necessary to ensure continued success in terms of performance and returns to investors), and the needs of individuals, who require some degree of control over their working lives (Carlopio, 1998).

Strategic Preparation

Strategic Knowledge and Awareness: New Business Models and Emerging Workplace and Technological Innovations

The first step in the process of implementing new technology and strategy or adopting new business models and products is to become aware that they exist and have some idea that they will help us achieve our organisation's strategic objectives. In this chapter we will focus on strategic knowledge and awareness as we explore some of the work done recently on new business models for the global enterprise and the technologies that enable them.

The Internet and Emerging Business Models

Porter (2001), the Harvard strategy guru (www.isc.hbs.edu), has come to several broad conclusions regarding the impact of the Internet on business. He suggests that although it appears to many that there are new rules of competition on the Internet, the final arbiter of business success is still the creation of true economic value. Porter uses his 'five forces' model (see p. 14 for more on Porter's five forces) to suggest that there are some consistent effects of the Internet on business. The Internet increases the threat of substitutes and can help make an industry more efficient. It tends to reduce the bargaining power of suppliers. It reduces differences among competitors, increases the number of competitors, increases pressures for price discounting and migrates competition towards price. It shifts bargaining power towards consumers and reduces switching costs. Finally, it reduces barriers to entry. While therefore acknowledging the effects of the Internet on many industries, Porter argues that it has not changed the fundamental rules. He concludes that the Internet is often not disruptive to existing industries or established companies and, while it does provide a new means of doing business for some, the fundamentals of competition remain unchanged.

According to Tapscott (2001), a business strategist and author (www.dontapscott.com), however, Porter is wrong about the Internet. Porter trivialises the Internet and considers it an incremental improvement in technology. Tapscott argues that the Internet is 'an unprecedented, powerful, universal communications medium' and as such represents something qualitatively new (Tapscott, 2001, p. 37). The Net is becoming ubiquitous and will soon connect every business and business function, as well as many humans, on the planet. The result, according to Tapscott, is the opportunity to network and form partnerships that are revolutionising the way we do business. It is established companies, not 'dot-coms', that are the main beneficiaries of the changes. For example, IBM and Dell have adopted various partnering models, basically focusing on their core capabilities and letting partners do the rest, and they have succeeded, whereas Apple, Digital Equipment Corporation (DEC), Prime and Data General stayed with the industrial age template of the vertically integrated corporation and have suffered the consequences.

In order to further our discussion of whether or not there really are any new business models that have been created because of the Internet, we must first be clear about what a business model is. According to Tapscott, the term 'business model' has been used ' in reference to everything from selling rocks online to a Vickery auction for financial services' and is often used synonymously with the term 'business strategy' (p. 37). The term business model is sometimes considered

> the totality of how the company selects its customers, defines and differentiates its offerings (or response), defines the tasks it will perform itself and those it will outsource, and features its resources, goes to market, creates utility for customers, and captures profits. It is the entire system for delivering utility to customers and turning a profit from that activity. (Tapscott, 2001, p. 37)

Tapscott, however, prefers the definition of a business model as the 'core architecture of a firm, specifically how it deploys all relevant resources (not just those within its corporate boundaries) to create differentiated value for customers' (p. 38).

Mahadevan (2000) agrees that the term business model is sometimes confusing. He said: 'consultants and practitioners have often resorted to using the term "business model" to describe a unique aspect of a particular

Internet business venture' (p. 56). He considers the business model to be a

> unique blend of three streams that are critical to the business. These include the value stream for the business partners and the buyers, the revenue stream, and the logistical stream. The value stream identifies the value proposition for the buyers, sellers, and the market makers and portals in an intranet context. The revenue stream is a plan for assuring revenue generation for the business. The logistical stream addresses various issues related to the design of the supply chain for the business. (pp. 58–9)

It is the term 'value for customers', from Tapscott's definition, that is key. Value for stakeholders is also central to Mahadevan's conceptualisation. A business model, then, may simply be defined as a way of making money, a way of doing business, a way of adding value for which someone is willing to pay.

One of the classic business models has been the vertically integrated corporation mentioned earlier in relation to Apple, DEC, Prime and Data General. This model suggests that the best way to make money is to control and add value to every step in the production chain. If you are an automobile manufacturer, for example, the suggested ideal business model was to own the mine producing iron ore, to own the steel mill as well as the parts manufacturer producing the necessary raw materials and components, and to own the design, assembly and sales functions. In this model, then, the way you make money is by vertically integrating up and down the production chain so you can optimise flow-through and inventories, capitalise on information flows, and reap the benefits of 'wholesale' costs throughout.

An idea that challenged this vertical integration model was specialisation. In a business model based on specialisation it is suggested that the way to make money is to focus on only one element in the production chain, such as parts manufacture or assembly. The value you add is by becoming 'world class' in a particular portion of the production chain. Other elements in the chain are provided by organisations that specialise in those areas. Rather than trying to do everything, therefore, you form partnerships and other cooperative arrangements with several 'world class' organisations. The question we need to answer in this section is what new business models, what new ways of doing business and what new ways of adding value for which someone is willing to pay have been enabled by the Internet, if any?

According to Turban, Lee, King and Chung (2000) the Internet enables four new business models related to electronic marketing.

1 Direct and indirect marketing. Direct marketing is when manufacturers advertise and distribute their products directly to customers via the Internet. This has been referred to as the process of disintermediation (Evans and Wurster, 2000). The more traditional situation is one in which a manufacturer's products are distributed through a third party. While it may be true that more organisations have taken advantage of disintermediation because of the ease of access to their customers provided by the Internet, any organisation that has a catalogue sales function has done the same thing. Cutting out the 'middle-person', therefore, is not new.

2 Electronic distributor. Electronic distributors take orders on-line and are responsible for order fulfilment and guarantee. This is the on-line equivalent of the more traditional situation and it is not at all new.

3 Electronic broker. An electronic broker introduces suppliers to customers and is not responsible for any other part of the transaction. Again, this is the on-line equivalent of the more traditional situation (e.g., an insurance broker) and is nothing new.

4 Auctions. While auctions have been viable business model for generations Turban *et al.* (2000) suggest 'the Internet provides an infrastructure for executing auctions much cheaper, with many more involved sellers and buyers' (p. 179). While it may be true that the Internet allows auctions to reach more people less expensively, auctions are not new either.

It seems, therefore, that Porter may be right. None of these models is actually new in that it was not possible for organisations to take advantage of them before the advent of the Internet.

Let us not yet give up hope of finding new Internet-based business models. Turban *et al.* (2000) also identified a number of supposedly new business models related to on-line publishing and knowledge dissemination (2000, pp. 183–4):

- the on-line archive approach is a digital archive, such as library catalogues and bibliographic databases
- the publishing intermediation approach can be thought of as an on-line directory for news services
- the dynamic approach is a 'just-in-time' approach wherein content is created in real time and transmitted on the fly in the format best suited to the user's location, tastes and preferences
- the Virtual University allows universities to offer classes worldwide
- on-line medical advice allows consultation with top medical experts

- management consulting on-line allows consulting firms to sell their accumulated expertise

All of these ways of doing business, of adding value and making money, also existed before the Internet. News media, libraries and universities, as well as medical and business advice, have all been available in many forms for a long while. So far, then, the Internet seems to be more of a new channel for communication than a revolutionary technology that has fundamentally changed the rules of business and stimulated the creation of new business models.

Glascoff (2001) discusses new Internet-based business models in the health-care system. He suggested that one successful model is the health-care portal, which, like the typical on-line portal such as Netscape, Yahoo! and AOL, usually generates its revenue from advertising. He also identified health-care connectivity sites, which make their revenue from transaction fees. Mahadevan (2000) similarly identified three new business structures/models: portals, market makers, and product/service providers:

- a portal engages primarily in building a community of consumers of information about products and services
- a market maker plays a role similar to that of a portal but also participates in a variety of ways to facilitate the business transaction that takes place between both buyer and supplier
- product/service providers deal directly with customers when it ultimately comes to the business transaction

Once again, I am not convinced any of these are really new. For example, various magazines, clubs and individuals have all primarily engaged in building a community of consumers of information about products and services.

Turban *et al.* (2000) also discussed supplier-oriented, buyer-oriented, and intermediary-oriented on-line marketplaces that have all been newly enabled by the Internet. In supplier-oriented marketplaces, both individual consumers and business buyers use the same supplier-provided marketplace. Examples of users of this model are Dell Computers, Cisco Systems and many electronic auctions. In buyer-oriented marketplaces a buyer opens an electronic market on its own server and invites potential suppliers to bid for the goods and services it needs. Large organisations can gain significant benefits from the competition this generates among would-be suppliers. In intermediary-oriented marketplaces a third party runs a marketplace where business buyers and sellers can meet. Examples such as ProcureNet and

Boeing's 'PART' fit in this category. None of these seem to have been impossible before the Internet. Many of these examples are variants on what we have discussed previously (e.g., supplier-oriented marketplaces are the same as electronic brokers or distributors) and are therefore not new.

Finally, Turban *et al.* (2000), like Tapscott (2001), discussed networking between business partners as an important new Internet-enabled business model. According to Tapscott, because of the unparalleled connectivity now available via the Internet, business-to-business partnering can occur in ways never before imaginable. Turban *et al.* (2000) discussed the Virtual Corporation as a new form of organisation composed of several business partners sharing costs and resources for the purpose of producing a product or service as an example of this new Internet-enabled networking. I find it hard to believe they have not heard of consortia, joint ventures, and the numerous other ways of pooling, allying and linking that have been discussed for years (cf. Kanter, 1989), and can think that networking or partnering are newly enabled via the Internet.

So, what is going on? If there are no new business models resulting from the advent of the Internet, what is all the hype about? Does the Internet really make no difference and change nothing? I think this would not be a justifiable conclusion.

My view is that while the Internet does not create any new business models, per se, it highlights how the digital transformation from atoms to bits that was heralded in 1995 by Negroponte in *Being Digital* plays itself out in the world of corporate strategy. Because of the speed and ubiquity of the Internet, for the first time we can see clearly the effects of this digital transformation.

What we can now see is that every business is an information business. From one perspective, an organisation's sole purpose is to capture, process and 'act on' information. Evans and Wurster (2000) illustrated how about one-third of the cost of health care in the USA (about $350 billion) is the cost of 'capturing, storing, processing, and retrieving information: patient records, cost accounting, and insurance claims. By that measure, health care is a larger information industry than the "information" industry' (p. 9). They discuss how a high-end Mercedes automobile has as much computer power in it as does a mid-range personal computer; how Toyota used information processes to gain competitive advantage in engineering, *kanban* (management of inventory), and quality control; how American Airlines used its 'SABRE' reservation system to achieve greater capacity utilisation; and how Wal-Mart exploited EDI (electronic data interchange) links with suppliers and the logistical technique of 'cross-docking' to dramatically increase inventory turns. The lessons are simple and powerful: information management, both high-tech and low-tech, is a key to competitive advantage. The value of the Internet is that it makes this more obvious and it enables us all to take advantage of these opportunities.

Blown to Bits: the Real Impact of the Internet

Evans and Wurster (2000), in their modern classic entitled *Blown to Bits*, suggested that while the supply chain and the value chain are useful ways to focus our attention on the important physical elements defining a business or industry, it is the information in the spaces between these elements that links them together. It is this linking information that really generates competitive potential and the value for which people are willing to pay. The Internet, the connectivity explosion and the growth of standards is melting the informational glue that holds the value and supply chains together. This is creating unprecedented strategic opportunities.

The fundamental catalyst for this explosion is the fact that the economics of 'physical things' is quite different from the economics of information: see Table 1.1.

The point is that as long as information and things are bound together, the difference in their economics is blurred and this suppresses value. There is always a trade-off, for example, between the informational imperative of wanting to provide the customer with unlimited choice, and the logistical imperative of wanting to minimise choice due to the associated costs (e.g., the cost of inventory and space to display and store many things). When information and things are blown apart, it unleashes the potential to create value in new ways. When selling compact discs and books, for example, separating the economics of things (e.g., independent warehouse delivery) and the economics of information (e.g., electronic search, and customer news and reviews) releases great value. The consumer obtains unlimited choice while the increased volumes eventually lead to more cost-effective warehousing and distribution.

Table 1.1   The economics of things and information

The economics of things	The economics of information
When sold, the owner gives it away.	Can be sold again and again.
High cost replication.	Low cost replication.
Has a location and unique legal jurisdiction.	Is nowhere and everywhere.
Subject to diminishing/increasing returns.	Has perfectly increasing returns.
Consistent with efficient markets.	Requires imperfect markets.

Source: Evans and Wurster (2000), pp. 15–16

The other trade-off popularised by Evans and Wurster is that between richness and reach. Reach is a simple concept. The more people who are exchanging information, the greater the reach. Richness is more complex. Richness can mean the amount of information (more information = more richness), the degree of customisation of that information (more customisation = more richness), the amount of interaction possible (more interaction = more richness), or the reliability, security, or currency of the information. The point is that usually with greater reach comes lower richness. When I communicate to a small group (low reach), I can communicate richer information (higher quantity, more customised, more interactive, more reliable, secure and current). When I communicate to larger numbers of people (higher reach), the amount of information, as well as its levels of customisation, interactivity, reliability, security and currency, are all likely to be lower (lower richness). This trade-off creates an unavoidable asymmetry of information. A 'seller' often knows more about a service or product (e.g., its quality, availability, what other providers charge) than the 'buyer'. The small number of sellers (low reach) have richer information. The general public (high reach) have less rich information. This asymmetry creates difficulties for both sellers and buyers (e.g., lack of trust, potential misunderstanding and loss of sales) and opportunities for others (e.g., third parties providing 'unbiased' information). The Internet is making information more widely available and redressing the asymmetry. Very rich information (a large amount that is customised, interactive, reliable, secure, and current) is becoming available to very large numbers of people (high reach).

This is creating the new role of 'navigator' for people (e.g., someone like me navigating through oceans of information and putting this book together), software (e.g., Quicken), evaluators (e.g., Consumer Reports), and search engines (e.g., Yahoo!) to provide that customised, unbiased information. It is also driving the deconstruction (i.e., the dismantling and reformulation) of traditional business strategies and business models. Intermediaries are especially vulnerable to this deconstruction. The Internet allows large numbers of people to get unbiased, rich information, and then to purchase directly from the source without intermediaries, such as retail outlets. Evans and Wurster coined the term 'disintermediation' to refer to this process. Intermediaries must re-invent themselves (e.g., by providing some information and/or service) or risk being made redundant. On-line trading has forced stockbrokers to drop prices and re-focus. Amazon.com has forced booksellers to add value to their physical outlets. Dell has forever changed the competitive landscape for personal computer sales. Automobile sales, retail banking, professional services, chemists, florists and many more have been, and will continue to be, affected. The Internet, then, is not necessarily fundamentally changing the rules of the game and therefore facilitating the creation of new business models; it is, however, providing many opportunities for people to adapt and combine existing business models creatively.

Evans and Wurster (2000) suggested that as long as connectivity continues, along with the adoption of common information standards, the richness–reach trade-off would continue to disappear. A consequence of this is that

> the channel choices for marketers, the inefficiencies of consumer search, the hierarchical structure of supply chains, the organizational pyramid, asymmetries of information, and the boundaries of the corporation itself will all be thrown into question. The competitive advantages that depend on them will be challenged. The business structures that had been shaped by them will fall apart. (p. 37)

This process of deconstruction has started and will continue to change the rules of engagement for the foreseeable future. To survive this deconstruction, you must manage for maximal opportunity, not for minimal risk. You must begin to deconstruct your own business or someone else will do it for you.

In order to summarise the effects of the Internet on business models, let us look at some work done by Kalakota and Robinson (1999). They suggested the Internet fundamentally alters the basic value proposition that can be offered to customers and allows us to redefine our business models. Kalakota and Robinson (1999, pp. 4–25) discussed eight rules of Internet-mediated business, as shown below:

1 Technology is sometimes the cause and driver of business strategy and is no longer an afterthought (cf. Yetton, Johnston and Craig, 1994). Much of what can be automated has been. We have squeezed a great deal of cost saving and speed from the application of computers to business. The Internet, however, can fundamentally alter the foundation of business. 'If any entity in the value chain begins to do business electronically, companies up and down that value chain must follow suit, or risk being substituted' (Kalakota and Robinson, 1999, p. 5). Executives who rely on IT (information technology) managers to make strategic technology decisions do so at their company's peril. Senior executives must be closely involved in technology decisions as they will increasingly determine our ability to position the organisation, to provide value to customers, to respond to market changes, and to guide innovation.

2 The ability to redesign the structure and flow of information creates more value than moving and manufacturing physical products. As Evans and

Wurster (2000) suggested, it is the information in the spaces between the physical elements in the production chain that really generates competitive potential and the value for which people are willing to pay. It is critical that managers start to separate the 'I' from the 'T' in 'IT'. Lumping information and technology together into 'IT' obscures real opportunity to add value and reduce costs. Organisations need information strategies and plans that are compatible with, but separate from, their technology strategies and plans.

3 An inability to modify a dominant, outdated business model leads to business failure. 'If it ain't broke, don't fit it' is the mantra of many. Unfortunately, by the time you are sure something is broken, it is too late to try and fix it. Most managers are focused on the short-term and are so busy fighting fires daily that they do not want to take the risk of, or even take the time to consider, abandoning their current strategy and toying with a tried and true business model. Today's market and technological realities, however, require organisations to be able to act and change 'at Internet speed' with alarming regularity. One response to this has been to outsource. First-generation outsourcing has focused on efficiency and cost-reductions. This new generation of outsourcing must be focused on pleasing customers. Networking and/or partnering can be used 'to quickly create reputation, economies of scale, cumulative learning, and preferred access to suppliers or channels' (Kalakota and Robinson, 1999, p. 8).

4 One outcome of the interconnectivity afforded by the Internet is the ability to combine business models and create flexible outsourcing alliances between companies that off-load costs and make customers ecstatic. 'This trend makes every market leader vulnerable' (Kalakota and Robinson, 1999, p. 9). The ability to disaggregate and reaggregate across the value chain with speed and ease is now a key organisational competence in many industries. Design, production and delivery are frequently best disaggregated, while the ability to reaggregate and form partnerships quickly, facilitated by the use of well-integrated enterprise software applications, is frequently necessary to create levels of value and service that entertains, links and binds happy customers.

5 The Internet is enabling companies to listen to their customers. We must become 'the cheapest', 'the most familiar' or 'the best' to survive (Kalakota and Robinson, 1999, p. 12). As Porter (1998) suggested, companies can successfully adopt a cost-based and/or a niche/differentiation-based strategy. The Targets, Big-Ws, Littlewoods and K-Marts of the world successfully compete on price. They use technology to ensure

superb inventory management and cost controls so they can be 'the cheapest'. The McDonald's, Coca-Colas and Yahoo!s compete by being 'the most familiar'. They use technology to help ensure a consistency of experience and they manage their brands. The Amazon.coms, Marks & Spencers and American Expresses of the world compete by being 'the best' products/services and by continually innovating. American Express' return protection policies, Marks & Spencer's 'refund money without question' and Amazon's low prices, unparalleled service (e.g., hard-to-find and out-of-print book location), third-party content (e.g., author interviews, pre-release information, reader reviews) and quick delivery all add up to an exceptional experience and exceptional value.

6 The Internet provides the opportunity to enhance the entire product/service-related experience. I recently used Microsoft's Expedia website (www.expedia.com) to find information about a family trip. It provided an astounding array of resources and tools including an interactive travel agent, a 'fare tracker', a 'mileage miner', flight status information, a hotel directory with maps, travel review and tips, weather information, 24-hour customer support and a very helpful currency converter (www.xe.net/ucc). The outcome was that not only did I find what I was looking for, but I found a couple of helpful related sites and services as well as some things I had not even dreamed of. It was an excellent 'holistic' experience.

7 Successful business models are increasingly using reconfigurable networks and partnerships to reduce costs and produce products and services that no one company can provide. For example, no one company can provide the type of full service experience found at sites such as Microsoft's Expedia. The level of coordination and integration required among members of these networks is high. This requires new types of managers, employees, organisational structures and cultures, measurement and reward systems, information systems, and so on.

8 The toughest task for managers is to align business strategies, processes and applications fast, right and all at once (Kalakota and Robinson, 1999, p. 24). Never before has the importance of integrating vision, strategy, cross-functional processes, software applications and IT infrastructure been greater. (We will discuss the importance of fitting all these elements together later in this chapter and in Chapter 2.) It requires remarkable leadership to be able to handle the big-picture vision and strategy, while appreciating the detailed implications of new technologies and managing the people-related change issues such as resistance, trust and communication.

Analysis of Competitive Forces

The first steps in the implementation process are formed around the processes of organisation and strategic level knowledge and awareness (Carlopio, 1998). In other words, people must become knowledgeable and aware of their external competitive environment and their internal strengths and needs. Externally, you must know your market and customers, your competitors (both existing and emerging), what your innovation and change options are, how government policy and regulation changes may affect you, and what new and emerging strategic opportunities and business models are available. Internally, you need to be aware of your core competencies, competitive advantages, strengths and weaknesses. From these internal and external scans comes the information you need to proceed to the next stage of the process (see Chapter 2). In this subsection, therefore, we will take a broad look at some of the competitive forces acting upon our organisations and our industries.

Analysis of External Competitive Forces

Your scans of the external environmental must be broad to ensure that you include a wide array of information. Harvard's Michael Porter (1998) suggested that five forces influence the competitive structure of an industry, which, in turn, affect the likely success of various strategic positions. Porter's five forces are supplier power, threat of substitutes, buyer power, threat of new end routs and the extent of rivalry in the industry.

The four strategic positions Porter identified are based on whether the strategic target is industry-wide or focused on a particular segment or niche within the industry, and whether the strategic advantage sought is that of low cost or some uniqueness perceived by the customer.

When taking a differentiated position, you are providing unique goods/services to everyone in the industry. When competing on overall cost leadership, you are also targeting everyone in the industry market, but your competitive advantage is your low cost, rather than your unique good/services. For example, Big-W in Australia, Wal-Mart in the USA and Tesco in the UK are positioned as overall cost leaders, competing on low cost across a full range of products, whereas Tandy Electronics stocks only a small portion of the products available in a narrower range of specialty electronics compared to Big-W and Wal-Mart, but is also targeting everyone in the market. Focused differentiation and focused low cost are the other two strategic positions. Ladies' fashion is focused on a particular market segment (i.e., ladies), but one ladies' fashion store (e.g., The Budget Shop or Fashion

Fair) might compete on low prices (i.e., focused low cost), whereas other stores (e.g., Laura Ashley, DKNY and Carla Zampati) might carry a unique line of more expensive designer clothes (i.e., focused differentiation).

Other models for the analysis of your external environment are less structured than Porter's. They simply ask you to consider lists of factors such as:

- social environment
- political-legal environment
- technological environment
- competitive environment
- consumer environment

or

- government
- economics
- customers
- markets
- technology
- media
- communications

The idea here is that you consider each of the factors and their likely impact on your industry and company. For example (see Figure 1.1 below), certain demographic changes (e.g., the ageing of the population) will probably affect tourism and the airline industry (e.g., more retired people with some disposable income and lots of time for leisure and holiday activities will travel more), as well as banking and telecommunications (e.g., increased need for financial planning advice and for people having access to their finds via their mobile phones).

Analysis and Selection of Internal Competitive Forces

Once the major external factors that are likely to impact on our organisations and industries have been identified, we have to consider our internal organisational environment (i.e., our organisation and its current capabilities, technology, structure, culture, processes, etc.) in order to determine which technologies, business models or other innovations will provide the

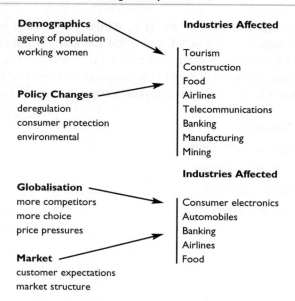

Figure 1.1   The effects of environmental factors on several industries

most strategic benefits. Although technology analysis, selection and decision are the focus of Chapters 2 and 3, in this section we will take a brief look at these topics at a strategic level.

There are many models that can be used when trying to assess your organisation's internal elements and functioning. For example, Galbraith and Nathanson (1978) suggested that there are five important internal elements: tasks, organisational structures, information and decision processes, reward systems and people.

Miles and Snow (1994) suggested a simpler model focusing on strategy, structure and processes (see Figure 1.2).

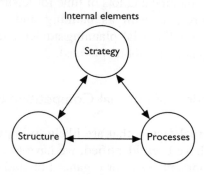

Figure 1.2   Important organisational elements according to Miles and Snow (1994)

While each of these models has its strengths and weaknesses, neither of them focuses on technology as a major element. In both models, technology would be considered as one of the organisational processes.

Since we are focused in this book on technology and strategy, I prefer a model that has come out of the Management in the 1990s Research Program conducted by a group of faculty members at the Massachusetts Institute of Technology (MIT) Sloan School of Management (Scott Morton, 1991). The MIT90s Framework (see Figure 1.3) illustrates that there are three core organisation culture-related issues (i.e., organisational structures and ways of working, management planning, power and control processes, and individual skills and roles) that are critical in the transformation process (Scott Morton, 1991, p. 21). These three core cultural elements are impacted by both strategy and technology.

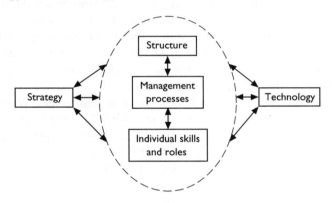

Figure 1.3   MIT90s framework

Source: Adapted from Scott Morton (1991)

Regardless of which model you prefer, the main point about these internal elements is that in order to be successful, an organisation's strategic direction, product market focus, and numerous internal elements must be aligned and fit together well in order to succeed (Fuchs, Mifflin, Miller and Whitney, 2000; Miles and Snow, 1994).

The concept of strategic alignment or strategic fit is critical to our conversation in this chapter: not only must you choose a successful strategic position (e.g., differentiation, or cost leadership, or high quality and service) in relation to the elements in your external environment (e.g., your customers and competitors), but you must also arrange your internal components (e.g., technology, structure, culture, staff, capabilities) so that you can deliver on that external position. For example, if your external positioning has you in a low-cost strategy, your technology and internal structures, culture, processes, systems and so on must be focused on delivering low cost. If this alignment exists, you are 'in fit' and you will be more successful than if your position was low cost, but your technology, systems, processes and procedures were designed to deliver customised services at high cost.

Let us look at an excerpt from Fuchs *et al.* (2000, p. 119) to help illustrate this idea of strategic fit:

> A close look at Wausau reveals that it had positioned itself as a niche innovator that rapidly enters, exploits, and then exits the most attractive niches of the paper market. The firm's ambition was to be the nimblest and the most adept competitor in those specific corners of the market. Because of its closeness to its customers and its acute market intelligence systems, Wausau was able to find product areas and applications that were highly profitable and unserved by larger competitors. However, that considerable talent would have been of little use were it not for Wausau's unparalleled flexibility of operations and its incredible speed in tooling up for new products and markets. The firm's fine strategic relationships with suppliers and subcontractors gave it the leanest and most flexible operations in the business. And its excellent distribution capabilities helped to keep the company responsive to its customers. Wausau's success, however, was not due to its astute positioning, nor its manufacturing flexibility, nor its market scanning and distribution capabilities, nor its strategic alliances. It was due to all of these things and to the exacting alignment that Wausau had effected among them in pursuit of its incisive strategy of niche innovation.

In other words, it was their internal alignment or fit that enabled them to deliver on their external positioning. In fact, fit is so important that even though Wausau is small, compared to its competitors within an industry that favours size, it 'has created more shareholder value in the last decade than any of its competitors and by 1997 had outperformed the industry average by a staggering 700%' (Fuchs *et al.*, 2000, p. 119). Fuchs *et al.* (2000) also studied several firms in the US retail industry (e.g., The Gap, a clothing retailer and Home Depot, a hardware retailer) and found that retailers who were more in fit, on average, had greater returns on capital. Their data revealed that organisations with moderate internal fit had about 50 per cent higher returns on capital than the average, while those with very good fit had almost twice the industry average return on capital (Fuchs *et al.*, 2000, p. 122).

In summary, strategic fit is important because it:

- supports the strategy and prevents leakage of efforts
- facilitates superior performance

- provides barriers to imitation
- simplifies complex organisational interdependence
- shapes desired behaviour
- reduces coordination costs

Business Models and Intangible Assets

To conclude this first chapter, and our look at new business models and strategy, let us consider the work of Boulton, Libert and Samek (2000) who discussed business models not by trying to develop a taxonomy of them, but by considering the role of intangible assets such as relationships, knowledge, people, brands and systems in delivering value in new ways. They suggested: 'the companies that successfully combine and leveraged these intangible assets in the creation of their business models are the same companies that are creating the most value for their stakeholders ... every company needs to create a business model that links combinations of assets to value creation' (p. 31). This is related to the topic of strategic fit and alignment. Boulton, Libert and Samek are suggesting that ultimate success comes not from the ability to make the most of just one or two assets, but from the ability to optimise all assets that make up a business model. The alignment and fit of these assets is key. Boulton, Libert and Samek proposed five strategies 'designed both to improve the value of specific assets as they interact, and also to enhance overall portfolio value, the value of the business' (2000, pp. 33–4), as listed below:

1 Strategy 1: Build. An organisation can build assets and develop or create new sources of value in many ways. Creating a new research and development process, for example, creates a source of value. Leveraging existing information and on-line customers by asking them to provide information when they visit a website is another example of building assets and creating new sources of value for which someone may be willing to pay.

2 Strategy 2: Enhance. Training is an example of enhancing the knowledge assets of individuals. As new technology and combinations of business models emerge, it is the organisations with employees who can leverage new technologies that will create and maintain competitive advantage.

3 Strategy 3: Connect. By connecting a company's discreet assets significant value can be added. Connecting the sales and service function available from one product or service line to another is the way to significantly add value. Connecting and integrating sources of information, islands of

technology and different elements within a production and/or a value chain are ways to create significant value.

4 Strategy 4: Convert. Assets can be converted to perform a different function or fill a different need: for example, convert your customers to salespeople and invest in technology and systems that are flexible and can be easily modified and converted when needed.

5 Strategy 5: Block. Boulton, Libert and Samek's (2000) fifth strategy is more defensive than the previous four. When blocking, an organisation uses an asset to make it more expensive and/or complicated for a competitor to build an asset portfolio that can successfully compete: for example, guaranteeing a high level of quality, value, and performance inhibits potential competitors from entering the marketplace. Locking-in suppliers and distributors, via technology or contract, is another example. First movers with new technology may also gain some competitive advantages as others may find it harder to break into a market and/or catch-up.

CHAPTER 2

Matching and Selection: Analysis of Workplace and Technological Innovation

In this chapter, we focus on the second stage of the strategic implementation process wherein we consider the knowledge we gained in the first stage (Chapter 1) and begin to sort through it as we form favourable or unfavourable attitudes towards certain new technologies, business models and other innovations. In this chapter we will explore several concepts and tools for technology planning, analysis and selection as we continue to explore innovations that enable global enterprise. We will also cover scenario development and analysis related to new technology. This will allow us to look at technologies, business models and other innovations that are on the horizon and assess which of these will best fit within our organisations or determine what needs to be changed so these innovations will fit. Ultimately we will be better able to anticipate the emergence of strategic opportunities that will give our organisations a competitive advantage. This initial sorting of options and opportunities is critical as we prepare to make our strategic decisions (covered in Chapter 3).

Scenario Planning, Development and Analysis

Successful scenario planning requires a balance between the known and the novel. Scenario planning can be used to help make people more aware of the external environment and, therefore, more prepared and motivated to adapt to and implement changes. If we think of the world of information as a pie chart, we can visualise something like Figure 2.1 to represent what we know.

Unfortunately, as we see in Figure 2.2, the amount of information that we do not know is twice as large.

The real problem, however, comes from the fact that the largest portion of all is what we do not even know that we do not know (see Figure 2.3).

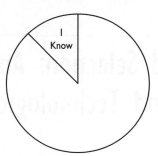

Figure 2.1

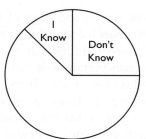

Figure 2.2

Figure 2.3

The things we do not even know that we do not know are our 'blind spots'. This is the area that really causes us trouble in our organisational and personal lives. Scenario planning can help us to shed some light on these blind spots and raise our awareness of what we do not even know that we do not know.

According to one of the scenario planning pioneers, De Geus (1988), a key to corporate longevity and success is the ability to adopt a survival mode when the business environment is turbulent, and to switch to a self-development mode when the pace of change is slow. De Geus suggested:

> While all companies learn, the crucial element is to be able to learn fast enough to sustain a competitive advantage. This is becoming increasingly important in today's fast-changing competitive world. The challenge facing companies is how to speed up the learning process among senior executives who have the power to act. Unfortunately, teaching is a very ineffective way to communicate information. Sometimes changing or suspending corporate rules can accelerate learning. A very effective learning tool, which can be described as a form of game playing, is developing 'what if' scenarios and planning responses to them.

According to Carol Humenick, Chief Executive Officer (CEO) of the $190 million Atlantic Credit Union in Newtown Square, PA (Humenick, 2000, p. 14):

> Implementing effective scenario planning, and incorporating staff into that process, has helped our credit union plan better for the future and achieve stronger staff and leader buy-in. The benefit to members is that we are more focused on building our electronic delivery systems and making anytime, anywhere access a reality.

They have implemented Internet banking, an enhanced audio-response system, a 24-hour loan application system, and have updated their debit card to real time, all in the same year. Humenick (2000) related how scenario planning enabled Atlantic Credit Union to be better prepared for when they were approached by three local credit unions regarding possible mergers. 'We were clear and all in sync,' Humenick said. 'We continue to learn that scenario planning is a more participatory technique than traditional planning methods. It is also a team building experience.'

Technology, Scenarios and Strategy*

Scenarios can provide valuable insights when trying to interpret change and can be used to help create successful strategic responses. Broadly, the process allows us to address the following type of questions:

---

*This section is adapted from Fahey (2000), van der Heijden (1996) and Devinney (1999).

1 What are some of the ways in which our competitive marketplace might evolve over the next three (or 5, 10, 20) years?

2 Precisely how might some hypothetical situation come about, step by step?

3 What are the factors, or forces, of change that are most likely to influence how our industry or marketplace will evolve?

4 What alternatives exist, for each factor, at each step, for preventing, diverting, or facilitating the process of change?

5 What are our best strategic and organisational options given these various potential 'futures'?

In most organisations, budget forecasting and strategic planning is extrapolated from the past and present. The questions above suggest as an ideal that strategy should be guided by an understanding of the future, not just the past or the present. Of course, this is not really possible to do with any certainty. Even though we can never be sure that we will 'get it right', we can still benefit from the process. As long as we can accurately anticipate trends, this will greatly facilitate our efforts. It is better to be vaguely right than precisely wrong.

According to Fahey (2000), scenarios serve two strategy-related purposes. First, they allow managers to anticipate a range of potential futures and to learn from them before they occur. Second, scenarios enable managers to consider what to do if each future actually materialised. Consider the examples in the following extracts (adapted from Fahey, 2000):

In the early 1990s, a large U.S. manufacturing company initiated a detailed analysis of its strategy alternatives in the European continent, from the Atlantic to Russia. It developed three scenarios, each detailing a distinct evolution of the widely anticipated political and economic integration of Western European countries and varying degrees of economic and political development in Central European countries. The three end-states were: (1) a buoyant economy with strong political leadership in all regions; (2) an economy with varying levels of economic buoyancy and political leadership and cooperation; and (3) a generally depressed economy with major political entanglements.

Each scenario suggested a variety of product opportunities, potential acquisitions of local competitors and suppliers, and ways to manage the company's European manufacturing, marketing, sales and logistics activities. These scenarios, along with the strategy and organizational implications derived from them, clearly demonstrated to the U.S. and European managers that the market opportunities in the late 1990s for the European continent far surpassed anything the company had previously envisioned. Also, the company's operations options were not nearly so narrow as indicated by the analysis that had led to its current European strategic plan.

As the European managers reflected on the Europe portrayed in each end-state, they identified which strategy and organizational alternatives they would pursue, how they would do so and what the results might be. They realized, perhaps for the first time, that significant opportunities would exist for their products, even if it took Central and Eastern Europe many years to reach the economic and social stability of Western Europe.

By engaging in scenario learning, a company can take the process of scenario creation a step further. Scenario learning shifts the emphasis from developing and refining the end-states, plots and logics to assessing the scenarios for their insights into the opportunities that may take the company in unexpected strategic directions or the lurking threats that could derail the current strategy. This is how we want to use the process. We are here concerned with identifying major trends in technology and business that will help us decide what changes to make and when.

Fahey (2000) suggested six principles that can help you get the most from the scenario learning process:

1 Scenarios only have value to the extent that they inform decision-makers and influence decision-making.

2 Scenarios only add value to decision-making when managers and others use them to systematically shape questions about the present and the future, and to guide how they go about answering them.

3 In each step of developing scenarios, the emphasis must be on identifying, challenging and refining the substance of managers' mind-set and knowledge.

4 Alternative projections about a given future must challenge managers' current mental models by creating tension among ideas, hypotheses, perspectives and assumptions.

5 The dialogue and discussion spawned by the consideration of alternative futures directly affects managers' tacit knowledge.

6 Scenarios should generate indicators that allow managers to track how the future is evolving. In this way, the learning induced by scenarios never ends.

The Scenario Learning Process

The outcome of the scenario development process is a set of descriptive narratives that present hypothetical, but plausible, alternative projections of a specific part of the future for the purpose of focusing attention on causal processes and decision points (Fahey, 2000; Kahn and Wiener, 1967). A set of scenarios could focus on macro issues regarding the implications of socio-economic and political forces and/or on more micro issues that detail how a particular industry might evolve, or how a set of technologies might interact over the next five years. Of course five people can develop scenarios quickly in a single two-hour session, or they can be developed after hundreds of hours of research and iterative development by scores of people. Scenarios can be focused on the short term, medium term and/or long term.

A *five-step scenario learning method*
The broad steps in the scenario learning process are as follows:

- definition of the field of analysis
- identification of critical influencing factors or drivers of success
- design of alternative scenarios
- derivation of adequate, scenario-supported strategic responses
- generation of flexible back-step strategies and responses

In the next few sections, I will take the perspective of an internal or external consultant trying to help a client through the outlined process. There are a number of group discussion and decision-making techniques that can be used to help at various stages in this process: see the Appendix to this chapter.

Step 1: Defining the field of analysis   The first task is to determine what is strategically important to the client. It is not always easy to bring to the surface and to articulate the strategic agenda. Some organisations have well-defined and articulated strategic plans and objectives; others have less well-developed strategies; still others have strategies that emerge and operate implicitly, in 'the background'.

It is essential to have a strategic positioning statement, or what some people refer to as the central business idea, articulated before proceeding with scenario development. This is what we discussed in Chapter 1 when we analysed our internal and external environments. This positioning statement acts as a 'base-line' and allows you to determine if any changes from the present course are needed. Strategic positioning statements are, for example, 'We will be the highest quality producer of X' or 'We are the lowest cost provider of service Y' or 'We help keep people healthy.' They help define the strategic field of analysis, as organisations cannot be all things to all people. Our strategic positioning statement articulates what we will and will not do or provide. It sets out how we position our company, products and/or service in relation to a number of factors (e.g., the competition, price, level of quality). It encapsulates our value proposition to customers, answering the question, 'What it is that customers actually purchase from us?'

Take, for example, the Swedish furniture company Ikea. They position themselves as providing style at low cost. They target 'young' furniture buyers, baby boomers who have money, children and little time. Their traditional competitors implicitly positioned themselves as providing full service and maximal customisation at various price-points. Ikea keep manufacturing costs down by employing modular designs, long-term suppliers and 'kit' packaging. They organise their stores in a 'walk through' pattern and use customer self-selection, self-assembly and self-delivery. This keep costs down and provides for immediate customer gratification. The services they do provide are highly valued by their target group (e.g., child-care, easy parking and parcel pick-up). Their competitors use floor samples and manufacturers' books containing pictures and fabric samples. They require vast floor-space, and provide maximal customisation which results in slower delivery and higher costs. It should be obvious how Ikea has clearly positioned itself differently from the Henredon Furnitures, Bowlings and Harvey Normans of the world.

Once you know what the current strategic agenda of the organisation is, you have one of the two critical components you need before you can actually start developing scenarios.

Step 2: Identify critical influencing factors and the drivers of success In order to focus the scenarios on issues that will significantly impact the organisation, it is important to have an appreciation of the critical factors driving the success, or failure, of the organisation. A classic SWOT (i.e., strengths, weaknesses, opportunities and threats) analysis can be used at this point. For example, the profits of organisations in the oil industry are particularly sensitive to global oil prices. Government departments are sensitive to ministerial party allegiance. Highly regulated industries would be particularly sensitive to changes in government policy. For example, the auto industry is, and will continue to be, affected by government regulations regarding exhaust emissions and the price of fuel. Many service industries are dependent on, and therefore especially sensitive to, changes in certain demographic elements (e.g., changes to the age or gender mix) of a population. For example, as the baby boomers have aged, they have had, and will continue to have, an impact on the finance industry and the entertainment industry. Many external issues within broad categories such as social, political, legal, technological, competitive, government, regulatory, economic, customers, markets, media, communications and consumer must be considered. Also, Porter's (1998) 'five forces' should be considered (i.e., the extent of historical competitive rivalry, the relative power of suppliers and buyers, and the threat of potential substitutes and new entrants).

When looking for potential influencing factors, be sure to consider the following types of issue:

- the social environment: education, values, demographics, entertainment, information, virtual communities, health care, welfare, telecommuting
- the technological environment: nanotechnology, telecommunications, robotics, computing (hardware and software), cable, television, multimedia, the Internet, standardisation, open systems, software development issues, information superhighways
- the political environment: infrastructures, laws, regulations, stability, tax
- the economic environment: interest rates, currency differentials, cash availability, banking stability, stock markets, world economy, distribution of funds, money markets
- the consumer environment: standard of living, options proliferation, global competition, supply and demand issues, distribution channels, emerging markets
- the business environment: large versus small firm size, economies of scale versus economies of scope, regional versus global
- energy: fuel cells, battery technology, solar, oil, gas, fusion
- materials technology: ceramics, composite metals, superconductors
- the global environment: pollution, emerging nations, war, natural resource availability (e.g., water, food, therapeutic food, clean air), recycling
- the medical environment: cloning, genetics, life-spans, super-drugs, super-bugs
- transportation: hybrid-fuel cars, hydrogen-power, intelligent highways, automated transportation systems

Another way to get at this information is to address the following types of question (adapted from Fahey, 2000):

1 What forces are shaping our current and emerging technical and business environment?

2 What new driving forces (i.e., new technologies) are emerging that might impact us, our competitors and/or our markets?

3 Which forces might not change much over the scenario period?

4 Which of these forces have more uncertainty associated with them?

5 How might these forces interact to give rise to one future rather than another?

Consider the following excerpt from Fahey (2000) as he refers to the questions above:

> These questions almost always lead to a discussion about the content and state of your company's knowledge – what it knows and doesn't know. Consider the following example: as part of identifying the driving forces that might affect a range of potential scenarios, a consumer foods company described a variety of demographic, leisure and social-value trends. Members of the scenario team began to collect and organize data and information pertaining to changes in geographic mobility patterns, income distribution across age groups, home ownership, eating out practices, and attitudes toward dieting and physical fitness.
>
> When they developed knowledge of these trends, they began to question whether some forces were as significant as they had previously thought. For example, they questioned whether average household income was as important a driving force as the company had historically believed. They also began to develop an appreciation for the power of social-value change as a driver of eating habits, the types of food eaten and where food is consumed.
>
> Thus, even in the early stages of outlining tentative plots and end states, those involved begin to develop a sense of how various trends and patterns in the external world connect to each other, and how they could affect the opportunities facing the company or its ability to pursue them.

Since we are particularly interested in technology in this book, be sure to consider the following types of technology question (adapted from Merrick, 1999):

1 Which technologies does your organisation use presently?

2 What are the costs and benefits of these technologies?

3 What are their strengths and weaknesses?

4 How can you improve your competitive position by maximising the use of existing technologies?

5 How are competitors using technology?

6 What technical, marketing, and management strategies do competitors use to get the most from technology?

7 What lessons can you learn from competitors' success or failure with these technologies?

8 What can you learn from your own experiences with new technologies?

To assess new technologies' effect on the market, try asking these questions.

1 What technologies are meeting or will soon be able to meet business plan goals?

2 What do competitors offer and how successful have they been?

3 What costs and benefits could technology bring?

4 What standards for success should you subject each technology to?

Next we have a trend extrapolation question:

1 How will a certain trend or parameter, continue into the future? For example, if the number of Internet users increases X per cent per year, there should be a certain number of Web users in three years.

Here are some useful substitution questions:

1 What new technologies are being, or could be, substituted for existing functions? For example, how could Internet delivery of information substitute for your current response?

2 What can we learn from the history of preceding technology introductions? Did recent technologies replace or augment preceding technologies? How quickly did they become profitable? What determined their acceptance in the market?

Use the past to predict the future:

1 What has happened in the last ten years in our industry that can help us understand how change might take place in the future?

2 Two years ago, what technologies were on the horizon? Which ones did you consider and how did they turn out? Did you make your technology decisions for the right reasons? Were they successful?

Of all the questions we can ask, the question of how technologies and processes will interact and influence each other is one of the most interesting, the most difficult and the most important. It is these unforeseen interactions that provide the most paradigm-shattering changes in our world. You can use the cross-impact question to compare the results of any of the preceding questions:

1 How will this technology interact with other processes and technologies? A cross-impact analysis looks at technologies in groups of two or more. For example, how will global communication technology affect production and/or your human resource information systems? Assume you install an information kiosk and begin transactions on the Internet. How will these developments affect your service levels and profitability?

When asking for this type of information, it is not unusual to find great diversity, and even conflict, among the perspectives of people from different backgrounds and functional areas. There are a number of group discussion and decision-making techniques that can be used to help in this process. For more information see the tools in the Appendix. This diverse information is critical to scenario planning as it highlights various assumptions about what drives success. These factors form the core of our next step.

Step 3: Design alternative scenarios   Our task at this point is to ask ourselves the key paradigm-shattering question (adapted from a Joel Barker video entitled *Paradigm Principles*, 1995): 'What is it, that is "impossible" today, that if it could happen, would fundamentally and radically alter our market, our company, our industry and our business environment?' Recall the pie charts from earlier in this chapter: it is what we do not know that we do not know that gets us in trouble. By asking this paradigm-shattering question we force ourselves to consider the fact that some of what seems 'impossible' today will actually happen. It also forces us to consider what our response options will be. It is important to remind ourselves that in the year 1400 it was impossible to sail around the world because it was flat. Only a little over 100 years ago, powered air flight and submarines were unknown. Electricity, telephone, radio, and computers were all unknown not too long ago. In the world of technology, impossible does not usually last forever.

Scenarios generally emerge from the following types of 'what if' question:

1 What if over 75 per cent of sales in our industry are generated over the Internet by 2010? What if a new technology makes the Internet irrelevant?

2 What if two of our main competitors form a strategic alliance?

3 What if a new form of service provides the same value we do at significantly lower costs?

4 What if a niche player enters our marketplace and takes 80 per cent of the business of our 20 most profitable customers?

According to Fahey (2000), in order to construct a scenario, you must structure and develop four interrelated elements: (1) an end-state, (2) a plot, (3) driving forces and (4) logics. The end-state refers to the state of the world at the end of the scenario period. For example, you must detail what the marketplace might look like in five years. These details may include the types of strategy various competitors would develop, how different classes of customers would interact with their suppliers, and how customers would obtain information about products, services and alternative suppliers. For a particular end-state to occur, specific plot details would have to take place. That is where the driving forces from the previous step come in. We must identify the events, decisions and actions of various factors that would cause a plot to evolve in a particular way. For example, you may have to determine the social, political, market and technological developments that would have to occur, and in what sequence, in order to reach a certain end-state; that is why the driving forces are so critical. If you are especially vulnerable to variation in the price of crude oil, and the prices drops or raises dramatically, this will have a major impact on your business. The fourth element of a scenario is a rationale for why a particular plot and associated end-state would unfold. For example, you will need a logic or rationale for why new e-business technologies would emerge; why new rivals would be able to enter the market; why specific customers would behave in certain ways; and what would motivate governmental policy-makers to enact certain laws, rules or regulations.

Each scenario developed must provide a distinctly different projection of a specific future. If a set of scenarios is developed around a narrow range of factors, the strategy alternatives and related choices across the scenarios will be too similar to be of value. One way around this is to construct two forced scenarios, one with all positive outcomes and one with all negative outcomes. Some pitfalls to watch out for include the following:

- generating too many alternative scenarios
- getting bogged down in overly elaborate and complex plots
- trying to complete the process quickly
- confusing scenarios with actual forecasts

- generating alternative scenarios on a single external variable leading to not enough variation in the scenarios

It is not easy to develop coherent and meaningful scenarios. Consider the following scenario examples (they may help you to develop some of your own). First, here are three post and one present scenario (adapted from Molitor, 1999):

1 Agricultural Age: this was premised on wresting sustenance and livelihoods from the land. US jobs peaked in this sector during the 1880s.
2 Industrial Era: this concentrated on mass production of fabricated products. US jobs peaked in this sector during the 1920s.
3 Service Era: here, undertakings involved employing the skills of third party providers to render specialised expertise the consumer could not perform so well by him/herself. Jobs peaked in this sector during the mid-1950s.
4 Information Era: at this stage, technologies rely on intellect and knowledge that educate, entertain, and manage human affairs. US employment in this sector has been dominant since the late 1970s. Attention currently is riveted on knowledge and education made possible by communication and computer linchpins. Still reaching toward its zenith, this current wave of technology has relatively few remaining years of dominance – possibly as few as 20 years – prior to being supplanted by another surging wave of economic importance.

Below are listed five potential futures (adapted from Molitor, 1999):

1 The leisure time era dominated by hospitality, recreation and entertainment. Leisure time pursuits have been a part of human activity from the very outset. The change about to be fully felt occurs when 'free time' dominates total individual lifetime activity.
2 The life sciences era dominated by technologies such as bio-technology, genetics, cloning, genetic engineering, transgenics, and 'pharming'. Their theoretical underpinnings can be traced back more than a century. The pace began to accelerate with the human genome project, and it reached a dramatic turning point with the cloning of Dolly.
3 The mega-materials era in which quantum mechanics, particle physics, nano-technologies, isotopes/allotropes/chirality, superconductors and microscopic imaging systems constitute the major core technologies. This sector began to take off with the development of plastics, bulletproof Kevlar, ceramic engineering, high-strength alloys, composites, silicon,

super-alloys, high-temperature superconductors, crystallography, cryo-
genics, semiconductors, time/temperature/pressure variable materials,
and designer materials.

4 The new atomic age in which thermonuclear fusion, lasers, and hydrogen
   and helium isotopes constitute the key technologies upon which almost
   every energy-dependent activity will depend. The predominance of these
   technologies looms ever closer as finite fossil fuels – first petroleum, then
   natural gas, and finally coal – are depleted. This era reaches its apex
   a century or more into the future. Roots of coming change, however,
   belong far back in time. This phase was born with 'splitting the atom'.
   Early experiments soon led to atomic fission, followed by development
   of thermonuclear explosives. Breakthroughs essential to harnessing
   fusion centre on advances in magnetohydrodynamics, laser-induced
   implosion, and quantum physics.

5 The new space age wherein astrophysics, cosmology, spacecraft development, space exploration, space travel, and resource gathering from
   space form pivotal activities propelling this stage of development.
   Beginnings for this sector can be traced back to gunpowder and rocket
   developments from over 2,000 years ago. Second World War rockets and
   jet aircraft accelerated the pace. Sputnik, spy satellites, manned space
   missions, extra-planetary probes and telescopic arrays that pierce the
   outermost limits of the universe are among the activities adding to the
   conquest of space.

*Step 4: Derivation of adequate, scenario-supported strategic responses and
Step 5: Generation of flexible back-step strategies and responses*   These
last two steps are totally unique to each situation and organisation.
Depending upon your business and industry, you must consider the implications of the various scenarios developed and determine what your strategic options are. Also, is it important to consider what the key indicators are
which would signal that one or another potential future is becoming more,
or less, likely. This is critical as these key indicators act as your signals to
initiate certain activities and make strategic responses.

At this point, it is important to review our current assumptions, expectations and strategies in light of each of the newly developed scenarios. We
must consider the following types of issue:

1 Where are the gaps and potential disasters?
2 How can we respond in each of the alternative futures?
3 How will the relevant stakeholders behave in each of the scenarios?

4 What is the range of uncertainty of the variables of interest?

5 Would our current strategy succeed in each future? If not, what strategy might succeed in each future?

6 Is there a robust strategy that would succeed in each potential future?

7 What are the crucial differences between our current strategy and the potentially successful strategies?

8 What would we have to do to develop and execute that strategy?

Some potential problems of which you should be aware include the following:

1 The prediction urge: the search for too much accuracy and no uncertainty.

2 The simplification urge: aversion to complexity.

3 Illusory expertise: specialists and experts are not necessarily the best forecasters. In fact, real breakthroughs and paradigm shifts usually come from 'outsiders' who do not know any better. People who do not already have preconceived ideas about what is, and is not, possible can think more broadly and creatively.

4 Sloppy execution:

   (a) poor selection of participants;

   (b) superficial analysis of responses;

   (c) poor design or process.

5 Overselling the technique: the process, even when done extremely well, is not perfect.

6 Typical problems of group dynamics:

   (a) group-think (the bandwagon effect);

   (b) influence effects of 'the boss' or other opinion leaders.

Technology Planning, Analysis and Selection

So far we have been dealing with broad issues and concepts in order to gain the background information necessary to make a decision regarding what technology, business model or other innovation we want to implement. In this section we will drastically narrow our perspective and focus specifically on the technology planning, analysis and selection process.

Strategic technology choices are necessarily made with a certain degree of uncertainty. The following tools are intended to help you make better decisions by providing structures to help organise and focus your thinking.

There are several 'analyses' we will conduct at this point. The first has four parts that look at the potential impacts and likelihood of success, the levels of familiarity with markets and technology, technology capabilities and technology life cycles. The second focuses more on risk (i.e., identifying, projecting, estimating, assessing, managing and monitoring risk). Of course, you can compare any variables that seem to make sense to you; the analyses provided here are merely examples.

Depending upon your strategy and your available resources, you can make any number of 'good' decisions, so there are no suggested outcomes for these analyses. Have a number of potential choices of technologies, business models or new products/services in your mind, and 'plot' them to get a visual representation of your total portfolio of choices. Adapt and change the analyses and analysis tools as you see fit (be sure to be able to justify your changes).

According to Stevens (2000), a key to successful IT portfolio management is the ability to hook-up strategic directions to real time and projected resource constraints of the organisation. Stevens (2000) stated:

> The theoretical portfolio is almost never completed. Good resource management will actually help get the portfolio done, and in a large engineering organization, this can only be achieved with some assistance from IT.
>
> Information technology can play key roles in three stages of portfolio management: at the strategic stage when the mix of projects is determined and prioritized to optimize return, minimize cost, and minimize risk; when individual projects are being moved from idea generation to commercialization and beyond; and as management determines optimal deployment of resources and manpower across an entire portfolio of projects.

At this strategic stage, risk analysis helps the portfolio manager better understand the potential risks and rewards involved.

Exhibit 2.1   Application exercise: technology portfolio choice tools*

Assessment 1   Potential impact and likelihood of success

The ultimate winners are those technologies with a high potential impact and a high probability of success. Unfortunately, in the real world, there is sometimes a trade-off between the two.

It is easy to see that those in the lower left of the matrix (Figure 2.4) should be abandoned. They are not likely to succeed and, even if they do, they will have little competitive impact. They are not worth your time.

---

* All four tools discussed in this exhibit have been adapted from work done by Professor T. Devinney, Australian Graduate School of Management, Sydney, Australia.

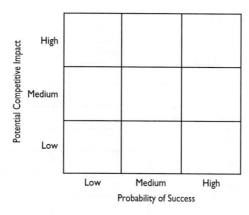

Figure 2.4

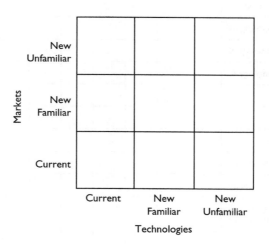

Figure 2.5

The more difficult choices are related to those that fall either 'in the middle', or those in the upper left (i.e., big impact with low probability of success) and lower right (i.e., likely to succeed with low impact). Choices regarding these types of technologies must be made depending upon your strategic positioning.

Assessment 2   Levels of familiarity with markets and technology

Two important elements to consider when assessing the overall likelihood of the success of a new product or service, for example, are the levels of familiarity with the new market and the associated technology (see Figure 2.5). Obviously, working simultaneously with a new market and a new technology is more risky than when you are familiar with one or both.

Technology CAPABILITY Audit							
	Current Position of Self/Competitors						Actionable Comments
	−3	−2	−1	1	2	3	
Organisational Resources							
Financial capability							
Production capacity							
Distribution capability							
Laboratory equipment							
Technological Expertise							
Technology A							
Technology B							
Technology C							
Human Resources							
Creative ability							
Innovative capacity							
Knowledge of key technologies							
Communication between divisions							
Past Performance							
Technology targets							
Commercial targets							
R&D cost control							
Programme dates							

Technology LIFE CYCLE Analysis

Technologies	Current Criticality (High, Med, Low)	Current Position of Self/Competitors						Future Critically (H, M, L)	Actionable Comments
		−3	−2	−1	1	2	3		
Current core (short term)									
Complementary (medium term)									
Peripheral									
Emerging (longer term)									
Obsolete									

Assessments 3 and 4

The next two tools are a technology capability audit and a technology life cycle analysis. Please remember that you can adapt and change these tools to suit your particular needs and circumstances. In the examples provided below, this analysis can be carried out for your organisation and/or for one or more of your competitors.

In the technology capability audit, you are asked to consider various resources or capabilities (e.g., human, technical, organisational) and to rate them on a scale that ranges from 3 to -3. Your perceptions of the relative strengths of you and your competitors across these variables should be an important factor in your technology decision-making process.

In the technology life cycle analysis, you are asked to list and consider your, and/or your competitors', technologies, as well as the current and future importance of these technologies. Your investments in technology should be considered across time horizons. A balanced portfolio will have projects in each of the three horizons: short term, medium term and long term, with correspondingly higher levels of risk associated with longer horizons.

Risk Analysis

Most information technology (IT) expenditure involves risk. The level of risk can vary significantly, however. For example, a non-critical project that is three months late will not make any significant impact on the business, but the failure of a critical project designed to add a new product or service to the business in a competitive market could have a significant impact on the balance sheet. Such a project failure could have catastrophic consequences if it were linked to other critical initiatives.

According to Abrahami (1999), in order to help ensure IT project success, we must be aware of the following:

- understand what risks are
- know how to quantify and qualify risks
- know how to mitigate and manage risks
- know how to avoid risks wherever possible, as prevention is more cost effective
- understand risk trade-off with risk management costs and returns
- have adequate project management, risk/impact analysis and change management skills
- integrated IT and business contingency plans

There are at least three distinct, well-known activities involved in an implementation project risk analysis (cf. Greenfield, 1998; Roman, 1986;

Walker, 2001): (1) risk identification, (2) risk projection, estimation, assessment, and (3) risk management and monitoring.

Risk identification

It is impossible to eliminate risk. There are, however, at least two things that you can do to try to manage risk: the first is to try to identify all the obvious risks and put plans in place to address them, and the second is to expect and plan for the unexpected. In other words, 'We expect that "X" could be a problem, so let's do something about it now.' We also know that there will be other problems that we cannot as yet even dream of, so we will put a process in place to identify them and to deal with them as they arise. We know problems will occur. We simply do not know what they will be as yet.

There are three categories of risk types that are traditionally considered:

1 Project risks: budgetary, schedule, personnel, organisation, resource, customer/user, complexity, size, structure.

2 Technical risks: design, interfacing, verification, maintenance problems, specification ambiguity, technical uncertainty.

3 Business risks: market risks, such as users not needing the innovation or technology; management risks, such as the innovation or technical change not solving the initial problem, not fitting within the culture which is too resistant to change, or not getting senior management commitment and support; and budget risks, such as the actual costs far exceeding anything anyone expected.

It is a fourth type of risk – social-psychological risk – that is usually not considered but with which we are most concerned in this book. When we are considering risk regarding the social side of the implementation of innovation and technical change, we are asking broad questions such as:

1 In the future, what non-technical factors might cause the project to go off the rails?

2 Who is 'for' or 'against' the project?

3 How will the project, and the changes it will bring about, affect existing power dynamics and relationships?

4 Who needs to be involved in the project design and implementation process?

5 How will changes (e.g., in requirements, people, power relations, structures, external environment, other parts of the organisation, our customers, technology) affect the project's timeliness and overall success?

6 What processes should we use to manage social-psychological risk? For example, what methods and tools should we use to manage risk, how many people should be involved and when, how big should the project be, how many iterations through the design, test, modify cycle should we count on?

Risk projection, estimation and assessment

In this analysis, we try to assess the likelihood that the risk is real as well as the consequences of the associated problems. Consider the nature, the scope and the timing of the risks. For example, you may be 100 per cent sure that a certain individual will be unhappy with the changes and will do everything to stop you. If that person has very limited influence, however, this is of less consequence. On the other hand, if it is 50 per cent likely that your senior management sponsor will leave the organisation in the next three months and the project will fail without this support, you will need a plan in place to manage that risk. For example, you should be working on identifying and capturing a secondary sponsor or two and/or getting project success embedded in many peoples' key performance indicators or performance assessment criteria.

Basically there are four degrees of risk severity (Roman, 1986):

- critical risk, which is unacceptable (e.g., catastrophic)
- potentially critical risk, which is acceptable on its own, but not in combination with other risks
- probably non-critical risk, which is acceptable but must occur in combination with other risks to lead to an unacceptable result
- non-critical risk, which is inconsequential

The levels of risk involved in the implementation of innovation and technical change vary with a host of organisational and technical factors such as the size, scope, and level of complexity of the changes.

When assessing the level of risk consider the following four broad questions:

1 How structured is the project? Do users know exactly what they want and what the innovation or new technology should do for them? Do those responsible for implementation know exactly how they will go about delivering the innovation or new technology? Do the implementers, developers and users agree on these points? If the answers are generally 'Yes', then the project is well structured. If the answers are generally

'No', then the project is more risky. Less structured implementation projects will (1) take longer, (2) require more iterations through the design, develop, test, and re-design cycle, (3) present those involved with greater dissatisfaction and reason for disharmony, and (4) be more likely to fail (i.e., they are more risky).

2 How experienced are those involved with the innovation or new technology being implemented? If you are moving to a team-based work organisation, and no one involved has ever worked in this type of organisation or has ever moved an organisation to a team-based work structure before, your level of risk is high. If you are implementing a new management information system in a division when people in other divisions have already done so, the risk is less as you have personal and organisational experience from which to draw.

3 How large is the project? Quite simply, the larger the project, the more complex the project, the more risky the project is likely to be. That is why pilot testing and prototyping are discussed and recommended as part of the initial and detailed implementation planning activities.

4 How big is the gap between 'where you are' and 'where you want to be'? If you have no technology in your firm, and you are planning to implement a state-of-the-art computer-based system to fully automate all aspects of your business, the gap between 'where you are' and 'where you want to be' is large, and so is your level of risk. Likewise, if you are a hierarchically organised firm, with strong centralised power relationships, and you are planning to move to empowered work-teams, the gap and your level of risk are high.

Risk management and monitoring

As just mentioned, if you know there is a fair chance that something will happen that can profoundly, negatively impact on your project, your task is to develop a plan of action to deal with it. For example, if you know that your organisation has a history of initially approving projects, and then losing interest, you must address this first and put in place mechanisms to ensure that people will not lose interest (e.g., put the necessary money aside up-front). There are four broad risk management strategies you should consider (cf. Greenfield, 1998; Roman, 1986), and see also the Risk Identification Tool (Exhibit 2.2):

1 Assumption or hedging. This strategy is to recognise risks inherent in the project and to build in contingencies by padding estimates of resources. You therefore assume risks and plan to deal with them as best you can by

	Areas of Impact										
	Profits	Quality	Business Continuity	Delivery	Occupational Health and Safety	Legal	Liability				
Sources of Risk								Likelihood of Consequences Occurring	Exposure Frequency	Rank

Legend:
Likelihood – Impossible, Practically Impossible, Conceivable, Remotely Possible, Unusual/Possible, Very Likely, Almost Certain, Will Happen.
Frequency of Exposure – Never, Very Rare, Rare, Unusual, Occasional, Frequently, Continuous, Always.

hedging. You can also hedge by putting in place multiple approaches toward the same goal, thus increasing your chances of success. If you need to get certain senior managers on side, you may hedge by giving them some information to read, having a peer talk to them and sending in a consultant to discuss the issues as well.

2 Transfer or exchange. When you manage risk via transfer, you exchange an unknown risk for a known risk that is more acceptable. For example, if going with a new piece of technology is simply too risky, you may decide to choose a well-known technology that is not optimal. In this case, you are exchanging the unacceptable technical risk for a known technical compromise that is acceptable.

3 Avoidance or evasion. Risk avoidance is done via planning. The goal is to avoid risks by conducting extensive planning, determining risks and

Once you have identified as many of the risks as you can, you now need to determine the mitigation approaches and who might assist you in preparing to address those risks. Use the following matrix to rank and formalise the risk mitigation plan.

Risk Treatment

Risk Description	Person Responsible	Rank of Risk	Proposed Risk Treatment Summary

possible consequences, and choosing alternatives that avoid recognised problems.

4 Reduction. This is the 'catch-all' category. You can broadly reduce risks by allocating enough time, money, people, efforts, and so on. You can also reduce risks by attempting to manage them in a variety of ways. For example, if there is the risk that some key senior manager will not support the project at a crucial decision meeting, you try to manage that person and get them on side prior to the meeting. If you need a specific person on the project team that is not available currently, you explore ways to free up that person and to interest them in the project.

Exhibit 2.2   Risk identification tool*

Risk identification is concerned with identifying those risks which may threaten the success of the solution. Such risks can be monetary or technical.

---

*Ross Seabrook and the team working on the Innovation Implementer project at Honeywell Australia from 1998 to 1999 originally developed this tool.

The following framework is a basic guide and a more detailed approach can be found in the Australia/New Zealand Standard AS/NZS 4360:1995.

Appendix: Tools for Scenario Learning

There are four tools outlined in this section.

1 The Nominal Group Technique.
2 The Delphi Technique.
3 The Stepladder Technique.
4 Edward De Bono's (1985) Six Thinking Hats.

The Nominal Group Technique (NGT)

The NGT is designed to help all team members participate and express opinions while still building team consensus. In this technique, seven to ten individuals are brought together to participate in a structured exercise that includes the following steps:

1 Team members are asked what they think of the issue (e.g., company's strategic position, value proposition or business idea or the strengths, weaknesses, opportunities, threats). Individual team members write down their answers, thoughts and ideas silently and independently. This is the nominal (non-interacting) phase.
2 Each team member (one at a time, in round-robin fashion) presents an idea to the group. As each idea is offered, it is summarised and recorded on a whiteboard or wall chart, without discussion of its merits.
3 A discussion is held in which all ideas are clarified and evaluated. The merits of the ideas are considered. Ideas are merged, eliminated, expanded and modified.
4 Individuals vote on each idea. This voting may involve a rating of the proposals, a rank ordering, the selection of the top two or three ideas, or the division of 10 points among alternatives. The group's output can then be prioritised or a single 'answer' can be reached by pooling the votes or rankings into a single preferred alternative.
5 A revised list of the best ideas is presented to team members for discussion. If a consensus emerges, the team is finished. If not, the procedure returns to Step 2 and continues through more rounds until the best ideas are identified and agreement is reached.

The nominal group technique is good as it avoids many of the potential problems of group decision-making: for example, a decision can be reached in a reasonable amount of time without being greatly influenced by the leader's preferred position. Perhaps the strongest drawback of NGT is its high degree of structure. As a result, the group may tend to limit its discussion to a single and often highly focused issue.

The Delphi Technique

Another technique for capitalising on a group's resources, while avoiding several possible disadvantages of relying on group decision-making processes, was developed by the Rand Corporation. This approach, called the Delphi Technique, is similar to NGT in several respects, but also differs significantly in that the decision-makers never actually meet. The steps in the Delphi technique are as follows.

1 Select a group of individuals who possess the appropriate expertise (e.g., forecasting technical trends or the effects of certain innovation or technical change options).

2 Survey the experts for their opinions via a mailed questionnaire.

3 Analyse and distil the experts' responses.

4 Mail the summarised results of the survey to the experts and request that they respond once again to a questionnaire. If one expert's opinion sharply differs from the rest, he or she may be asked to provide a rationale. This rationale could then be forwarded to the other participants.

5 After this process is repeated several times, the experts usually achieve a consensus. If not, the responses can be pooled to determine a most preferred view.

The Delphi technique has a number of advantages and disadvantages. Its greatest advantage is that it avoids many of the biases and obstacles associated with interacting groups (i.e., groups where the members meet face-to-face). It has also been shown to generate fairly useful information and high-quality solutions. A strong disadvantage stems from the amount of time it takes to complete the entire Delphi process. It rarely takes less than several weeks, and often as long as several months. Clearly, urgent problems cannot be solved in this manner. Finally, like NGT, the Delphi Technique follows a highly structured format. As a result, it does not offer much flexibility if conditions change and obviously, since respondents never meet face-to-face, social interaction and free dialogue are lost.

The Stepladder Technique

This method provides a solution to the problem of unequal participation in groups. The Stepladder Technique is intended to improve group interaction by structuring the entry of group members into a core group. Initially a small group of two members, for example, work on a problem. Then the third member joins the core group and presents his or her preliminary suggestions for solving the same problem. Next the member's presentation is followed by a three-person discussion. Each additional member, fourth, fifth and so on joins the expanding core group and presents his or her preliminary solutions; at each step there is a discussion.

The technique has four requirements. First, each member is given sufficient time to think about the task before entering the core group. Second, the new member must make a preliminary solution presentation before hearing the core group's ideas. Third, sufficient time is allocated to discuss the problem as each person is added. Fourth, the final decision occurs only after the entire group is formed. These steps make it difficult for a member to 'hide' in the group. Research on the effectiveness of the stepladder technique has shown that stepladder groups produce higher-quality decisions than conventional groups.

Edward De Bono's (1985) Six Thinking Hats

One of the difficulties in many groups is that people cannot break out of their existing mind-sets or see problems from different perspectives. De Bono's method helps structure the thinking process so that people deliberately take on various 'hats' or ways of looking at a problem or situation. His six 'hats' are:

- white hat: facts and figures
- red hat: emotions and feelings
- black hat: critical thinking (what is wrong with it?)
- yellow hat: speculative-positive (optimism, benefits, positive what if?)
- green hat: creative and lateral
- blue hat: control of thinking (thinking about thinking, examine the process)

You can use these 'hats' in different ways. One way is to assign different individuals to wear a certain coloured hat (i.e., take on a certain perspective) for a certain period of time. Another way is to have the entire group shift from 'hat' to 'hat'. Either way, it allows a fuller range of perspectives to be brought to bear on a problem or situation.

Decision: Technology Acquisition

In this chapter we consider strategic/organisation level decision-making, business case analysis and cost justification. After becoming aware of what technology and innovations are available (Chapter 1), and what we need based on our scenario and technological innovation planning, analysis and selection (Chapter 2), a decision is made to adopt certain new technologies or innovations and to reject others.

1 'We have to revise then more fully integrate our method of handling forecast production and sales information into our MRP system.'
2 'We must do some business process re-engineering to cut costs and streamline our order and payment processes.'
3 'We will move to a flatter, more organic, organisation structure.'
4 'We will implement a new management information system.'

If you are considering a new strategy or technology, this is the point at which you make many of the decisions regarding the 'who, what, where, how, why and when' related to the proposed new direction(s).

Strategy and Decision-Making: Is Strategy Decided Upon, or Does it Emerge?

One of the most important and 'violent' debates in this field is between those who think that strategy is something that can be analysed, planned, designed and decided upon, and those who believe that strategy is something that emerges over time and can only be recognised 'after the fact'. This debate is critical for those of us concerned with implementation and strategic decision-making. If strategy emerges, there is nothing to analyse, sort, decide upon and implement. If we can design strategy, then it needs to be analysed, decided upon and implemented.

Before we explore this debate a bit further, I'll give you my 'answer'. Strategy is both emergent and designed. Anyone who does not analyse, plan, design, decide upon and implement strategies, tactics and goals is completely directionless and adrift. Equally importantly, however, anyone who analyses their situation, makes strategic decisions, designs a strategic response, develops a plan and then rigidly sticks to it at all costs, with no account taken of how the environment reacts and changes, is unrealistic and inflexible. Either of these two extremes, in the absence of its opposite counter-balancing force, leads to eventual disaster. The implications for implementation are that we need to focus on implementing the strategies, decisions and plans that are made, and also stay mindful of the fact that as we do this, we must leave the organisation, and its systems and people, open to implementing the inevitable 'corrections' and modifications that emerge and need to be made along the way.

Strategy is Designed versus Strategy is Emergent

The rationalist, planning, design school of thought is the implicit model used in most of the strategic planning literature (epitomised by Michael Porter, *Competitive Strategy*, 1998). This is the view we explored in Chapters 1 and 2. In this view, we analysed the internal and external environment, developed and explored various strategic options, and rationally chose the 'best' new technology or strategic response given our analyses. After this, we go about implementing our decisions. According to Mintzberg and Lampel (1999), this prescriptive, design school of thought views strategy formation as achieving the essential fit between internal strengths and weaknesses, and external threats and opportunities. Senior management formulates clear, simple and unique strategies in a deliberate process of conscious thought. The process is assumed to be cerebral, formal and decomposable into distinct steps, delineated by checklists, and supported by techniques. According to van der Heijden (1996), one of the fathers of the scenario planning method, 'Managers love the rationalist school, it assigns power to them to determine the destiny of the organisation. However, they also realize that it does not always work very well' (p. vii). This later qualification seems to me to be an implicit recognition that strategy is simultaneously designed and emergent.

The evolutionist, emergent school of thought (epitomised by Henry Mintzberg) is much less palatable to managers. It takes the power away from managers, suggesting that strategy emerges over time, involves many people, and can only be understood in retrospect. This more descriptive perspective has, at its philosophical base, a belief in retrospective sense-making (Weick, 1979). The idea is that we can only understand something

as complex as strategy by looking backward in time and making sense of what has already happened, retrospectively. The emergent school of thought views strategy as a learning process in which, for example, new kinds of strategies emerge from collaborative contacts between organisations. It assumes that strategists can be found throughout the organisation, and that the strategy formulation and implementation processes intertwine (Mintzberg and Lampel, 1999).

In this emergent strategy view, it is suggested that instead of following a rational, linear strategy formulation and then an implementation process, managers should facilitate the involvement of many people, experiment with many options, encourage random, chaotic connections between people and ideas, and realise that the formulation and implementation processes are iterative. According to Mintzberg and Lampel (1999), a problem that results from this descriptive, emergent view of strategy is that managers can end up tangled in confusion, generating multiple contingencies and perspectives that inhibit implementation. This seems to me to indicate that, while this emergent view frees up the behaviour of those who follow the linear, design school and enables a fuller and richer view of the strategic process with more experimentation and innovation, without the discipline and clarity of the design school nothing actually gets done.

Strategic Decision-Making

Decision-making is one of the central activities of management and is a critical activity during this phase of the technology implementation process. The quality of the decisions made at this point will profoundly affect implementation success and the long-term viability of your organisation. Many decisions that have to be made during the process of implementing innovation and technical change will need to be made by groups. Group decision-making is different from making decisions on your own. In an effort to increase our decision-making effectiveness, both as individuals and in groups, information on three important issues is covered in this section: programmed versus non-programmed decisions, individual versus group decision-making, and some of the perceptual and judgemental factors affecting decision-making.

Programmed versus Non-Programmed Decisions

One way of distinguishing among decisions is in terms of whether they are fairly routine and well structured, or novel and poorly structured

(Vecchio, Hearn and Southey, 1996). Well-structured decisions can be understood, measured and actually 'programmed'. For example, when a clerk checks the on-hand inventory against a pre-established minimum standard this well-structured decision can be analysed, easily measured in terms of success or failure and pre-programmed. If inventory falls below the standard, the clerk knows it is time to order more stock.

Poorly structured decisions, on the other hand, are more ambiguous and frequently more difficult to make. With decisions that are unique and non-routine, taking a programmed approach is usually impossible. These non-programmed decisions frequently pertain to rare and unique situations that have a potentially significant effect on the organisation. For example, major technology planning issues are non-programmed decisions. How to acquire capital, whether or not to sell off unprofitable corporate divisions, and whether to launch a new product line or not, are other examples of organisational non-programmed decision issues. These types of decision afford the greatest opportunities for creativity (Vecchio, Hearn and Southey, 1996). They are more difficult to make, frequently have significant impact on the firm and are very likely to be made during the course of the implementation of innovation and technical change.

Individual versus Group Decision-Making

Most decisions have both individual and group components. It is, however, sometimes difficult to know when to involve others in the decision-making process and to what degree. By and large, research that has pitted individuals against groups has shown that groups will outperform individuals working in isolation. That is, the groups' solutions to problems are typically of higher quality than the average of the individuals' solutions. One interesting additional finding is that the best solitary worker may often outperform the group. In general, however, and for a variety of tasks, groups can be expected to outperform the vast majority of individuals who work alone.

Precisely why groups have an advantage over individuals has also been the subject of much research. One self-evident explanation is that groups can draw from a larger pool of information and abilities. By pooling these resources, the group gains access to a collection of knowledge that is greater than that of any single individual. This knowledge enables the group to reject obviously incorrect approaches and provides a check on the possibility of committing errors. Being in a group also tends to motivate and inspire group members. The stimulation of being in a social setting can enhance an individual's level of contribution. In addition, there are social rewards for making a significant contribution to a group's efforts. For example, praise, admiration, and feeling valuable to the group can be strong incentives for an individual to exert greater effort in accessing valid information and evaluating decision options. Finally, depending on the situation, it may be possible to divide a group's general assignment into smaller, more manageable tasks that can then be delegated to individual group members.

There are also several potential disadvantages to group decision-making. For example, highly cohesive groups sometimes encourage a restricted view of alternatives. This is sometimes referred to as groupthink. In this case, the group is too cohesive and will not let in any new information or perspectives that disagree with the group's line of thinking. Groups may also polarise towards extreme points of view if an appreciable element of risk is involved. In other words, group members tend to feed off of each other's fears or enthusiasm, and can make what researchers refer to as risky shifts and cautious shifts (cf. Bateman, Griffin and Rubinstein, 1987; Karan, Kerr, Murthy and Vinze, 1996). Another potential problem is that group decision-making tends to be much more costly than individual decision-making. Given the time and energy that meetings can consume, it is usually best to reserve group decision-making for more important decisions that require high-quality solutions. Group discussions can also give rise to hostility and conflict. This is especially likely when group members have divergent and strongly held opinions on alternative courses of action. In addition, decision-making in groups tends to be influenced by the relative status of group members. Thus, when a group member who possesses relatively little status offers an objectively good suggestion, it may be rejected. If a group member with high status offers the same suggestion, however, the likelihood of its being adopted is greatly increased.

Perceptual and Judgemental Factors Affecting Decision-Making

We now briefly turn to potential biases and problems of which we must become aware, in order to increase our decision-making effectiveness as individuals or in groups. Human beings are fallible information processors. It is not possible for us to perceive and process information completely objectively (cf. Bazerman, 2002). This is not a matter of opinion: it is psycho-physiological fact. For example, we all have a physical blind spot caused by the optic nerve entering our retina. In our everyday perceptions, however, this blind spot does not show up as our brain filters it out for us. There are a number of perceptual factors that affect decision-making and judgement of which we are also usually unaware.

1 Limited information-processing capability. Humans have a limited and biased information-processing capability. When we look at a room or out of our windows, we do not 'see' everything that is there. We attend to certain things and we 'ignore' others. Most people can read only a few hundred words per minute (Burns, 1996). There is a limit to the amount of information we can perceive via our hearing as well. There are times, therefore, that we may think we heard what someone said, but we misheard or misinterpreted it. There are times when we think we have all the information we need, but we merely have all the information we are willing and able to handle for the moment. 'Research … consistently shows that the human mind automatically discounts information that is inconsistent with what a person already believes or wants to believe and places disproportionate weight on information consistent with the person's beliefs and desires' (Bazerman and Loewenstein, 2001, p. 28). This can have a significant negative impact on our ability to make successful decisions.

2 Perceptual filters and biases. We have many filters and biases, and we frequently rely on ineffective decision-making heuristics (Bazerman, 2002). A decision-making heuristic is a guide or 'rule-of-thumb'. For example, we may use the 'rule' that our first choice is always best, or always wrong. Either way, this cannot possibly be a successful base for decision-making. When interviewing a potential new employee, we frequently see our perceptual filters and biases in action. Many of us dislike certain 'types' of people or expect that other 'types' do well in certain jobs. This can also have a significant negative impact on our ability to make successful decisions.

3 Changing perceptions. Our filters and perceptions change, via learning, experience, framing, attitude change and belief change. For example, the 'anchoring effect' is a consistent and powerful psycho-perceptual phenomenon (Ritov, 1996; Whyte and Sebenius, 1997). Here is how it works. If I am trying to sell you a car for £35,000, I get you to talk about and think about the price of a new home. That anchors you into thinking about hundreds-of-thousands of pounds. Now, when you think about the price of the car, it seems less than it would seem if I had anchored you to a lower amount related to the price of a new suit, for example. When we process information, therefore, we are frequently effected by the information we collect and by how and when it is presented to us. This also can have a negative affect on our decision-making success.

4 Inaccurate perceptions. Our perceptions are subject to factors of attention, stress and many perceptual phenomena. When we are tired or stressed, our judgement and perceptions can be impaired. If we are focusing on one part of a picture, we may not be able to see another part. Many of us have seen

examples of perceptual illusions. Once again, this illustrates how we need to remember that we are fallible information processors and must, therefore, take steps to minimise these potential biases.

As a result of these, and many other perceptual factors, our ability to judge accurately and objectively is sometimes affected. For example, research shows that our first impressions are frequently not accurate, and frequently long lasting. It is well established that we tend to remember best what we hear and see first and last. These are referred to as the primacy and recency effects (Bazerman, 2001). If a vital piece of information is presented in the middle of a series of useless facts, we may be likely to forget it.

Another example of how our judgement can be affected is related to our tendency to stereotype. Unfortunately, while stereotyping helps us to survive and to handle large amounts of data, it is consistently related to biases in performance ratings, attitudes and prejudices about people of different races, gender and nationalities (cf. Arvey and Campion, 1982; Barclay, 1999; Orster, 1994; Schmitt, 1976). Research illustrates that attitudinal, racial and gender similarity can affect evaluations. It has long been recognised that people tend to like those who are similar to themselves rather than people who are different (cf. Byrne, 1971). Studies consistently show that people like those who had attitudes similar to their own. The greater the similarity, the greater the attraction. In other words, you must be OK as you are just like me. Cross-cultural comparisons illustrate that this similarity-attraction principle is universal. Respondents from India, Japan and Mexico all showed the same patterns as 'Western' respondents.

If we want to make the best decisions regarding the implementation of new technology, it is vital that we are aware of these perceptual inadequacies and inconsistencies. Awareness is the first step on the road to minimising their effects. With this information as background, we can now turn to the topic of business case analysis. Most decisions regarding the strategic implementation of technology involve a business case.

Business Case Analysis

One of the most salient considerations at this stage in the strategic decision-making process is how to quantify the potential risks and benefits of the proposed changes. Whether we are implementing a new strategy, a new structure or some new technology, the topic of cost justification is critical to implementation success. This issue is frequently unmentioned in research and teaching in this area. This may be because these issues are considered the domain of accountants and financiers and not managers, or because these cost-justification issues have not been very well understood or explored until recently. I think it is a major mistake for change managers not to be exposed to several of the critical issues.

The business case is a tool that helps us make decisions about whether or not to buy a technology, which vendor to choose, and when to implement it. Business cases are generally designed to consider the question: 'What are the likely financial and other business consequences if we take an action or decision?'

The business case is not a budget, and it is not a management accounting report. A business case does, however, show expected cash-flow consequences of various actions over time as well as the methods and rationale that were used for quantifying benefits and costs. The business case also describes the overall impact of the proposal in terms that financially astute managers look for such as discounted cash flow, payback period, and internal rate of return. Critical success factors and contingencies also must be discussed in a business plan as this explains what must be managed or done in order for the predicted results to be achieved. Significant risks also should be codified along with indicators that would signal changes in predicted results.

While it is not possible to prescribe a single business case template that will be appropriate for all situations, the following is an example of some of the sections and subsections one might find in a typical business plan (adapted from *Business Case Essentials: A Guide to Structure and Content* by Marty J. Schmidt, 1999, www.solutionmatrix.com and *The New Dynamic Project Management* by Deborah Kezsbom and Catherine Edward, 2001, New York: Wiley):

Confidentiality statement

Title page
    title and subtitle
    address and author
    date
    subject
    purpose
    disclaimer

Executive summary
    business description
    current position and future outlook
    management and ownership
    funds sought and use
    financial summary

Introduction and overview

Assumptions and methods
    financial metrics
    assumptions
    scope and boundaries
    cost/benefit models
    data sources and methods
    full value figures versus incremental figures

Business impact
    financial model
    cash flow statement
    analysis of results
    non-financial results

Sensitivity, risks and contingencies
    sensitivity analysis
    risk analysis
    contingencies and dependencies

Conclusions and recommendations
    conclusions
    recommendations

A Cost Typology for Information Technology

Assessing the full costs of new technology is not as simple as accounting for the purchase of new equipment and software. A host of factors must be considered and many questions must be asked, including the following:

1 What are the committed costs of the proposed system, taking into account peripherals, essential supplies, insurance, maintenance contract, upgrades, software development, etc.?

2 What future costs are likely to be incurred on maintenance, extending and linking the equipment, and software development?

3 What costs are likely to be incurred in the redesign of other related systems or equipment to ensure compatibility?

4 What is the cost of any special environmental control equipment to ensure security and overcoming the potential damage from heat, humidity, dust, electrostatic interference, vibration and direct sunlight? What is the cost of any special workspace layout requirements (e.g., to reduce the

noise of printers) what is the cost to arrange appropriate decor, lighting, furniture, work services and storage space?

5 What are the space requirements for the new equipment, peripherals, support equipment, operators, power supply and cabling, air-conditioning, materials handling, and waste removal?

6 What happens if the system goes down? Are there costs incurred due to potential system unreliability? Are redundant systems necessary, for example? If so, what costs are involved?

Riel (1998) suggested that there are three categories of costs associated with information technology (IT) projects: (1) technological, (2) systems, and (3) support costs. Technological costs are mainly hardware related and can be considered as IT infrastructure. They include the cost of equipment but also other costs such as studies and technological obsolescence. Systems costs are more software related. They include costs from programming, designing, and operating the system that results from using the infrastructure. Support costs are indirect costs pertaining to maintenance of both hardware and software. Each of the costs is discussed in more detail below.

Technological costs of IT*

Cost of study of new IT   The first cost when contemplating IT projects is the technical and business analyses that are necessary to figure out what is needed in terms of technology. Such studies can be very complex and can take several weeks, if not months. To define the future IT system (computer equipment, peripherals and telecommunication links) that will be aligned with strategic objectives, a great deal of information may be needed. Outside help may be hired (vendors and consultants) and training may be required. The lack of proper analysis at this stage significantly increases the technological, implementation, and organisational risks of IT projects. Some firms do not properly perform the appropriate studies at the outset of IT projects. This mistake usually ends up in increased costs later, when IT systems need to be redefined, reprogrammed and redesigned. This means that there are opportunity costs with these studies; bypassing the planning stage may save money in the short term but will increase implementation costs (which are part of the support costs) later by a factor of two or three.

---

* This section is adapted from Riel (1998).

Cost of reach, range, and responsiveness One of the fundamental questions to ask about prospective IT projects is to define the desired IT platform. The IT platform depends on specific technological characteristics, namely reach, range and responsiveness:

- reach: the accessibility of business processes
- range: the level of information sharing
- responsiveness: the quality of service

The successful implementation of business policy with respect to technology depends on its ability to connect people together (reach) inside and outside the organisation, on its ability to provide complex interactions between people (range), and its ability to process information quickly and reliably (responsiveness). These are the main characteristics that emerge from information technologies.

An IT system that can only connect people within a department has less reach than one that connects organisations together. Likewise, an IT system that only transmits messages has less range than one that enables complex interactions between people. Obviously, these capabilities are expensive. IT projects are defined in terms of the incremental reach, range, and responsiveness the organisation is going to need to satisfy its strategic ambitions. Certain kinds of equipment will provide more reach and/or range. The added costs of such additional capabilities must be weighed against the additional benefits they generate.

Cost of technological risk Technological risk becomes business risk when a significant portion of the firm's cash flow is tied to information technology. Project failure is not unusual and the probability of less than expected levels of performance is high.

Opportunity costs of infrastructure reach, range, and responsiveness Inappropriate infrastructure system reach, range, and responsiveness are likely to end up in lost profitability: hence the presence of opportunity costs. Such costs come from lost businesses due to insufficient just-in-time processes, dysfunctional cross-functional coordination, and inefficient supplier–customer cooperation. IT projects should be aimed at reducing to the greatest extent possible the opportunity costs associated with these potential situations.

Opportunity costs due to equipment obsolescence Everyone knows from personal experience how a new personal computer rapidly becomes obsolete. IT suffers from the same phenomenon both in terms of hardware and software. Systems obsolescence can translate into higher operating costs, lower benefits and lower profits compared with competitors that have newer systems. Thus these costs comprise increased operational costs on the one hand, and lost business opportunities on the other hand. Obviously, such irreducible costs are not easily estimated but we can be certain that they exist.

Systems costs of IT

Programming costs   These costs can be particularly crucial for IT projects as they can rise rapidly. Programming is a tedious task and its productivity can be low, hence the presence of hidden costs. Moreover, what drives programming costs is not obvious, as research has shown.

Cost of study of new system   As with the technological cost of new IT studies, this cost is likely to be significant due to the complexity of the system to be defined. Information engineering may be required and this kind of analysis can also be lengthy and costly in terms of time and resources. Besides the system's specifications, the associated business process it supports may also have to be analysed concurrently as both IT and business processes are intimately linked.

Reorganisation, process redesign costs   IT projects influence the kind of business processes that will be present in the future within organisations. The problem is to find what kind of processes can liberate the potential of IT. This means that implementing IT probably implies redesigning business processes, which is not cheap since such changes are usually substantial.

Opportunity costs of system reach, range and responsiveness   As for infrastructure, inappropriate system reach, range and responsiveness are likely to end up in lost profitability: hence the presence of opportunity costs. Software must also be designed that goes along with the infrastructure.

Support costs of IT

Cost of maintenance   This cost is perhaps the most prominent in its class. Debugging is part of support and can be costly both in terms of time and resources. Much of this cost can be hidden due to insufficient planning or excessive systems complexity.

Table 3.1 lists the major categories and some examples within each. Costs are listed as tangible, irreducible or intangible. A tangible cost can be

Table 3.1

IT costs	Tangible costs	Irreducible	Intangible
Technological	• Cost of study of new IT • Infrastructure hardware costs of range, reach and responsiveness • Replacement and modification cost of incomparable equipment	• Opportunity cost of technological risk: probability of dysfunctional performance • Opportunity cost of equipment failure and obsolescence	• Costs due to thriving technological changes that cannot be assimilated fast enough • Opportunity cost of misalignment between technology and business strategy
Systems	• Cost of study of new system • Software costs • Programming design costs • Software costs of reach, range and responsiveness	• Opportunity cost of training, conversion, reorganisation, redesign • Opportunity cost associated with infrastructure reach, range, responsiveness • Cost of transition delays and errors	• Opportunity cost associated with reliability and effectiveness of decision-making
Support	• Cost of installing new equipment • Testing • Layout costs (furniture) • Costs of performance studies after implementation	• Opportunity cost of insufficient IT system study (redesigning, reprogramming, etc.) • Opportunity cost associated with working conditions/ergonomics • Opportunity cost due to intense debugging, improving system on behalf of users	• Opportunity cost associated with quality of work life

assessed because its measurable effect can be translated from operational to economic terms. An intangible cost, by definition, cannot be assessed in dollar terms because there is no operational (or quantifiable) measure of its effects in the organisation. However, this leaves another kind of cost that lies between the two extremes. Such costs are, as we could say, somewhat measurable. They are typically quantified through some operational performance measure, but require a special cost model to translate their physical effect into monetary terms. Their economic effect is either lost in some overhead pool or simply not recorded in accounting systems. Flexibility costs fall into that category. Because such costs are neither fully tangible nor completely intangible, they are referred to as irreducible (Riel, 1998).

Human Due Diligence

A part of the technology decision-making process and business plan analysis that frequently seems to be missing is a consideration of the people side of the equation. While financial and technical analyses provide critical information, they simply do not address the high-stakes human risks that many new technologies and business models generate. So important are people to the change process that Conner (1999) and Cooney (1999) recommend carrying out a people-oriented version of financial due diligence before any major changes are undertaken. In this view

> people have an elastic but limited ability to adapt to change and this adaptive capacity should be treated as a finite resource as valuable in its way as any financial capital. Before any major change such as a company-wide restructuring or the installation of a new information system, leaders need to survey their organization to see if their people have the willingness and/or ability to accommodate a major change. Designed to increase an organization's adaptation capacity, human due diligence involves looking at things like management's commitment to change and employees' knowledge of change dynamics. (Cooney, 1999)

Human due diligence brings structure and discipline to the people side of the business case analysis and decision-making process. Conner (1998, pp. 99–100) states:

> The term 'due diligence' refers to the investigation done before taking important actions. Most people are familiar with financial due diligence: critical information is gathered before key decisions are made and carried out. The term implies not a cursory review, but an extensive and comprehensive investigation of the issues and implications surrounding vitally important decisions. Instead of instinct or guesswork, a serious, rigorous effort is made to be sure the findings are as reliable as possible ... because most senior officers relate to the due diligence process as a serious matter, Human Due Diligence is the term high use with leaders who are serious about orchestrating the people aspects of implementation architecture. It brings to the people side of change the

same demanding meticulousness and painstaking thoroughness normally associated with financial due diligence. To the emotional, social, and cultural aspects of change, it applies the exactitude and attention to detail typically put to use during important acquisitions or vital technical matters that are under consideration.

Human due diligence has three primary components (Conner, 1998, pp. 106–7):

1 Collection: information is gathered about an organisation's overall capacity to prosper during prolonged ambiguity (i.e., enterprise and preparation) and/or its ability to successfully implement a specific project or series of projects (i.e., tactical implementation).
2 Planning: once proper due diligence information has been collected, it forms the basis for planning whatever actions are necessary to mitigate the problems or exploit the opportunities that were identified.
3 Action: based on the plans developed, the activities engaged in are intended to either increase adaptation capacity and/or reduce implementation demands.

The idea of readiness to change is critical to Conner and Cooney's concept of human due dilligence. Research has recently focused on the importance of readiness for change as a variable related to the success – or failure – of individual and organisational change efforts (Eby, Adams, Russell and Gaby, 2000; Stewart, 1994; Tucker *et al.*, 2000). Consider the following readiness for change quiz to help you get a sense of the concept.

Readiness for Change Assessment*

The left-hand column in the chart lists 17 key elements of change readiness. Rate your organisation on each item. Give three points for a high ranking ('We're good at this; I'm confident of our skills here'); two points for a medium score ('We're spotty here; we could use improvement or more experience'); and one point for a low score ('We've had problems with this; this is new to our organisation'). Be honest. Do not trust only your own perspective; ask others in the organisation, at all levels, to rate the company

---

* This tool is adapted from Stewart (1994).

too. It may help to have an outsider, or someone who can be more objective, do the assessment with you.

SPONSORSHIP: The sponsor of change is not necessarily its day-to-day leader; he or she is the visionary, chief cheerleader, and bill payer, the person with the power to help the team change when it meets resistance. Give three points – change will be easier – if sponsorship comes at a senior level; for example, the Chief Executive or Chief Operating Officer, or the head of an autonomous business unit. Weakest sponsors: mid-level executives or staff officers.	Rating:
LEADERSHIP: This means the day-to-day leadership – the people who call the meetings, set the goals, and work until midnight. Successful change is more likely if leadership is high level, has 'ownership' (i.e., direct responsibility for what's to be changed) and has clear business results in mind. Low-level leadership, or leadership that is not well connected throughout the organisation (across departments), or that comes from the staff is less likely to succeed and should be scored low.	Rating:
MOTIVATION: High points for a strong sense of urgency from senior management, which is shared by the rest of the company, and for a corporate culture that already emphasises continuous improvement. Negative: tradition-bound managers and workers, many of whom have been in their jobs for more than 15 years; a conservative culture that discourages risk taking.	Rating:
DIRECTION: Does senior management strongly believe that the future should look different from the present? How clear is management's picture of the future? Can management mobilise all relevant parties – employees, the board, customers, etc. – for action? High points for positive answers to those questions. If senior management thinks only minor change is needed, the likely outcome is no change at all; score yourself low.	Rating:
MEASUREMENTS: Three points if you already use performance measures of the sort encouraged by total quality management (defect rates, time to market, etc.) and if these express the economics of the business; two points if some measures exist but compensation and reward systems do not explicitly reinforce them; if you don't have measures in place or don't know what we're talking about, one point.	Rating:
ORGANISATIONAL CONTEXT: How does the change effort connect to other major activities in the organisation? (For example: does it dovetail with a continuing total quality management process? Does it fit with strategic actions such as acquisitions or new product lines?) Trouble lies ahead for a change effort that is isolated or if there are multiple change efforts whose relationships are not linked strategically.	Rating:
PROCESSES/FUNCTIONS: Major changes almost invariably require redesigning business processes that cut across functions such as purchasing, accounts payable, or marketing. If functional executives are rigidly turf conscious, change will be difficult. Give yourself more points the more willing they – and the organisation as a whole – are to change critical processes and sacrifice perks or power for the good of the group.	Rating:
COMPETITOR BENCHMARKING: Whether you are a leader in your industry or a laggard, give yourself points for a continuing programme that objectively compares your company's performance with that of competitors and systematically examines changes in your market. Give yourself one point if knowledge of competitors' abilities is primarily anecdotal.	Rating:
CUSTOMER FOCUS: The more everyone in the company is imbued with knowledge of customers the more likely that the organisation can agree to change to serve them better. Three points if everyone in the work force knows who his or her customers are, knows their needs, and has had direct contact with them. Take away points if that knowledge is confined to pockets of the organisation (sales and marketing, senior executives).	Rating:

REWARDS: Change is easier if managers and employees are rewarded for taking risks, being innovative, and looking for new solutions. Team-based rewards are better than rewards based solely on individual achievement. Reduce points if your company, like most, rewards continuity over change. If managers become heroes for making budget, they won't take risks even if you say you want them to. Also, if employees believe failure will be punished, reduce points.	Rating:
ORGANISATIONAL STRUCTURE: The best situation is a flexible organisation with little churn (i.e., reorganisations are rare and well received). Score yourself lower if you have a rigid structure that has been unchanged for more than five years or has undergone frequent reorganisation with little success; this may signal a cynical company culture which fights change by waiting it out.	Rating:
COMMUNICATION: A company will adapt to change most readily if it has many means of two-way communication that reach all levels of the organisation and that all employees use and understand. If communications media are few, often trashed unread, and almost exclusively one-way and top-down, change will be more difficult.	Rating:
ORGANISATIONAL HIERARCHY: The fewer levels of hierarchy and the fewer employee grade levels, the more likely an effort to change will succeed. A thick layer of middle management and staff not only slows decision-making but creates large numbers of people with the power to block change.	Rating:
PRIOR EXPERIENCE WITH CHANGE: Score three if the organisation has successfully implemented major changes in the recent past. Score one if there is no prior experience with major change or if change efforts failed or left a legacy of anger or resentment. Most companies will score two, acknowledging equivocal success in previous attempts to change.	Rating:
MORALE: Change is easier if employees enjoy working in the organisation and the level of individual responsibility is high. Signs of non-readiness to change: low team spirit, little voluntary extra effort, and mistrust. Look for two types of mistrust: between management and employees, and between or among departments.	Rating:
INNOVATION: Best situation: the company is always experimenting; new ideas are implemented with seemingly little effort; employees work across internal boundaries without much trouble. Bad signs: lots of red tape, multiple sign-offs required before new ideas are tried; employees must go through channels and are discouraged from working with colleagues from other departments or divisions.	Rating:
DECISION-MAKING: Rate yourself high if decisions are made quickly, taking into count a wide variety of suggestions; it is clear where decisions a made. Give yourself a low grade if decisions come slowly and are made by a mysterious 'them'; there is a lot of conflict during the process, and confusion and finger-pointing after decisions are announced.	Rating:
Now add together your 17 ratings for your total score	Total score:

If your total score is:

41–51: Your organisational readiness for change is high and implementing change is most likely to succeed. Focus resources on lagging factors (i.e., the items you rated with scores of 1 or 2) to accelerate the process.

28–40: Your organisational readiness for change is moderate and change is possible but may be difficult, especially if you have low scores in the first seven readiness dimensions. Bring those up to speed before attempting to implement large-scale change.

17–27: Your organisational readiness for change is low and implementing change will be very difficult. You may want to reconsider your change plans and focus instead on (1) building change readiness in the dimensions above, and (2) effecting change through 'skunkworks' or pilot programmes separate from the organisation at large.

Implementation

CHAPTER 4

Knowledge and Awareness

Communication for Change

Research and experience clearly illustrate that to ensure implementation success it is a change manager's responsibility to adopt a comprehensive communications strategy that includes the following types of activity (cf. Brimm and Murdock, 1998; O'Neill, 1999; Radosevich, 1999):

1 Develop a stated objective, or message, and stick with it. The vision communicated must be clear and the reasons for the change must also be communicated. Information must help people understand 'why' the changes are necessary in order to reduce anxiety and resistance to change. It is not easy for people to change. It is even harder to change when there is no clear incentive for doing so. Change managers must link individual success with corporate success. In other words, tell employees what is in it for them as well as how it relates to the business.

2 Get the information out to all employees, even those who were not directly affected by the new technology. Different audiences respond to different media, so be prepared to disseminate your message using a number of forms of communication. Devise a communications plan including options such as weekly meetings, an Intranet newsletter, and an internal television network to broadcast daily updates in public areas and e-mail broadcasts to keep everyone abreast of the project status and goals. Also consider various pilot-testing options such as user-trial sessions demonstrating the capabilities of the new technology.

3 Identify key champions, preferably first-line managers/direct supervisors, and use them to lead the changes where appropriate and to test internal communication strategies before going company-wide with them. People automatically distrust euphemism and jargon. When you say 'rightsizing', employees hear 'lay-offs'. When you say 'high performance partnerships', people hear 'more work, fewer resources'. If you are going to establish trust, say what you mean, and mean what you say. Speak to people in the language they understand. Use your champions to test your language. Finally, remember effective communication is a two-way

process. You must do as much listening as talking, and you must be prepared to deal with what you hear. Your champions can be your eyes and ears.

Remember National Culture

This is not a book on international management or cross-cultural communication, per se. I would, however, be remiss if I did not highlight at this point the fact that everything said here is culturally biased. Everything here is formed within the 'Western' cultural tradition. What may be considered polite behaviour and skilled communication by one group of people may not be considered so by another group. There are distinctly different norms regarding such issues as communication formality/informality, eye contact, hierarchy and power, vocal tone and the use of humour (cf. Hofstede, 1980, 1981, 1983, 1992; Trompenaars, and Hampden-Turner, 1998) across national cultures.

It is essential, when communicating in the global enterprise, always to be sensitive to cultural and diversity-related issues. This is especially important when communicating for change. For example, motivational and change-related communication frequently draws on metaphors and 'figures of speech' to conjure images of desired end-states and to affect the way people see the world (Akin and Palmer, 2000). We must be aware that referring to sports stars and historical figures, for example, may not be effective cross-culturally; what is considered as common sense and common knowledge in one place may not be considered so in another. For example, in the 'West' we would expect anyone with the least bit of common sense and education to know of people such as Winston Churchill and John F. Kennedy, as well as Shakespeare, Mozart and Picasso. People in the 'West', on the other hand, would have no idea who Yu Gong and Meng Jiangnu are. If you grew up in China you would have to be pretty unaware not to know that referring to Yu Gong alludes to how persistence pays off and that Meng Jiangnu displayed unparalleled loyalty. Most 'Westerners' would not know why President Sukarno is noteworthy (the first President of Indonesia), and would never have heard of Kohnosuke Matsushita (founder of the Matsushita Group), Michio Miyagi (famous Japanese composer/musician), Shikibu Murasaki (a female aristocrat born at the turn of the first millennium and one of the world's first novelists) or Basho Matsuo (known as the first great poet in the history of haiku). Similarly, in southern Europe, if you did not know that Salazar was a dictator who ruled Portugal for over 30 years throughout the 1940s, 1950s and 1960s, and that to refer to some one as 'Salazarista' means you consider them fascist, you would have to be pretty unaware. In northern Europe, in the Slavic speaking countries, everyone is taught in primary school that Cyrillus and Methode were brothers who invented, in the ninth century AD, the Cyrillic alphabet used in Slavic languages. Every Indian knows of Nehru not because of the Nehru jacket, but because of the man who wore it as Prime Minister from 1947 to 1964. In Pakistan, figures such as Mohammed Ali Jinnah (the founder of Pakistan), Allama Mohammed Iqbal (poet of the East) and Abdul Sattar Edhi (who runs orphanages, hospitals, schools and reform centres for drug addicts) are part-and-parcel of the national culture. The cultural relativity of our knowledge and values has a significant impact on our ability to communicate, relate and do business across cultures.

How Great Leaders Communicate

Research shows consistently that the ability to communicate effectively is one of the most critical skills in determining success across a wide array of organisational outcomes (cf. Carlopio, Andrewartha and Armstrong, 2001). One way to learn about successful communication for change is to study the words and deeds of proven leaders. In this section, therefore, we will focus on several leaders from outstanding companies: Jack Welch (General Electric, USA), Lawrence A. Bossidy (Allied Signal, USA), Bob Mansfield (Optus Communications, Australia), Percy Barnevik (Asea Brown Boveri, Swedish-Swiss) and Akio Morita (Sony, Japan). It is surprising how consistently these five very different individuals from different industries and nations speak about leadership and communication.

The Importance of Communication

Many people are afraid of change, technology, job loss and more. Because of this fear, in the absence of accurate and timely information, people will fill in the gaps and actually 'make things up'. People abhor an information vacuum. This is one reason why rumours start. Especially when leading change processes, leadership equals communication. Information is the lifeblood of an organisation. Every time two people who do not usually talk to each other do so, it creates new potential within the firm. When you want to stimulate change, you want information coursing through your organisation. Communication has been referred to as the very essence of a social system or an organisation. Communication is essential because the structure, effectiveness and scope of organisations are almost determined entirely by communication. If we take away communication, we would not have an organisation. Innovation and change means increased uncertainty, a lack of order, predictability and stability. Information and communication means decreased uncertainty. Therefore, the more we change, the more communication is necessary to counter the uncertainty.

According to Jack Welch from GE (Slater, 1994), the task of a great leader is to help transfer ideas, allocate resources, and then get out of the way. Managing less, controlling less, filtering less is better. Communicating more is a result and this is better. Welch says leaders must: 'make sure everyone in your business gets all the information required to make decisions. Managers control and complicate. Leaders facilitate and simplify.'

Bob Mansfield, while he was the Chief Executive Officer (CEO) of Optus Communications (a leading Australian telecommunications company), was quoted as saying:

> Optus's future success depends on its ability to communicate, both externally with customers and internally among its 4,000 staff. Communications is the glue that makes it all stick together, ... but you can't just talk about it and then go back into your office and shut your door. You really do have to walk the talk ... most communications among Optus's top management group is informal, with managers using electronic mail, discussing problems as they meet in corridors or dropping into colleagues' workspaces for a few minutes conversation. There is only one formal meeting of the top management team every two weeks, while Mansfield himself addresses and answers questions from staff on a national video hook-up every month. (Roberts, 1995)

Successful leaders, therefore, spend a great deal of their time talking to people. In fact, they seem to take every opportunity to discuss their vision and plans. Morita of Sony suggests one of the keys to successful leadership is a genuine, and sometimes bold and daring, focus on communication. For example, he made it a point to personally address all of the incoming graduates each year, and Morita used to have dinner with many young lower management employees almost every night and talk until late, to foster working relationships and to keep in touch. 'One reason we can maintain good relations with our employees is that they know how we feel about them' (Morita, 1987, p. 143).

Similarly, Bossidy from Allied Signal (*Harvard Business Review*, March–April 1995) says:

> besides talking to large groups, whenever I go to a location I host smaller, skip-level lunches, where I meet with groups of about 20 employees

without name tags and without their bosses. I think [this] combination ... gives [me] a pretty good handle on how people think about things.

Jack Welch from GE also suggests: real communication takes countless hours of eyeball to eyeball, back and forth. It means more listening than talking. It's not pronouncements on videotape; it's not announcements in the newspaper. It is human beings coming to see and accept things through a constant interactive process aimed at consensus. (*Harvard Business Review*, September–October 1989)

Leadership communication is so critical it seems successful leaders think it is better to over-communicate than it is to under-communicate. To be successful, according to Percy Barnevik, CEO of the Swedish-Swiss heavy industry firm Asea Brown Boveri (ABB), 'You don't inform. You over inform', as there is a strong tendency to be selective about sharing information. Barnevik states:

I have no illusions about how hard it is to communicate clearly and quickly with tens of thousands of people around the world. ABB has about 15,000 middle managers prowling around markets all over the world. If we in the executive committee could connect with all of them or even half of them and get them moving in roughly the same direction, we would be unstoppable. (*Harvard Business Review*, March–April, 1991)

Especially since most of us tend to under-communicate significantly, when we communicate enough, it may seem like more than it really is.

What to communicate
In terms of what to communicate, it is essential that leaders decide, and then clearly communicate, the direction and means of travelling from point A to point B. In other words, leaders must set a direction, goals, vision and so on, and then communicate this information to others clearly and continuously. These signals must be loud enough to be perceived above the background 'noise' of everyday business and information, and must be consistent enough to provide a unified field of vision.

Two illustrations of these principles can be seen in Welch at GE and Bossidy from Allied Signal. Bossidy said: 'In 1991, we were hemorrhaging cash. That was the issue that needed focus. I traveled all of the company with the same message and the same charts, over and over ... if we don't fix the cash problem, none of us is going to be around.' Keep it simple, Bossidy suggests; 'we're spending more than we're taking in. If you do that at home, there will be a day of reckoning' (*Harvard Business Review*, March–April 1995). According to Welch at GE, strategy has to evolve, not be etched in stone. Do not pursue a single, central idea. Set only a few clear general goals as business strategies and then get on with it. For example, Welch said GE was to be 'number one or number two' in every business they were involved in. He advocated expressing a vision, then letting employees implement it on their own (Slater, 1994). Welch reminds leaders, however, that their decisions and plans must not be too rigid or 'etched in stone'. Welch says leaders must

> Look reality in the eye and be ready and willing to change your agenda if what you have planned no longer seems to fit with the external environment. In deciding how to change your business, nothing should be sacred. Take a hard look at your overall business and decide as early as possible what needs fixing, what needs to be nurtured and what needs to be jettisoned. Face reality. Forget about excuses, problems and the circumstances that stop us. Define it and do it. (Slater, 1994)

How to communicate

Communication must not only be adequate in terms of quantity, it must be consistent and truthful. According to Bill Gates, 'Passionate leadership won't succeed if contradictory signals are sent' (Microsoft website www.microsoft.com, 12 September 1996). Truthful, honest communication requires trust. Fear is the greatest obstacle to honesty. People resist change and 'invest in' belief systems out of fear. Unresolved anger blocks listening and leads to passive-aggression and resistance. It becomes not safe to explore, challenge, change, risk, or be involved.

It is the responsibility of the leader of the enterprise to foster an atmosphere in which it is safe to challenge the status quo. Bossidy from Allied Signal suggests that leaders 'want to create an environment in which people will speak up. Every question is interesting and important. When I conduct interactive sessions, I don't walk out after three questions. I make it clear that I'm going to be there until the last question is asked.' Similarly, Morita suggests that a diversity of ideas and opinions is necessary for organisational health and success:

> When most Japanese companies talk about cooperation or consensus, it usually means the elimination of individuality. At our company we are challenged to bring our ideas out into the open. If they clash with others, so much the better, because out of it may come something good at a higher level. (Morita, 1987, p. 146)

According to Morita, if two people have exactly the same ideas on all sub-
jects, are totally cooperative and are never in conflict, then one of them
should resign as 'it would not be necessary for both of us to be in this company and receive a salary ... It is precisely because you and I have different
ideas that this company will run a smaller risk of making mistakes' (Morita,
1987, p. 147).

There are many potential modes of communication. Generally, written
forms of communication (e.g., memos, reports, e-mail) are better for long
or complicated messages. This gives people more time to process the information. Do not under-estimate the power of informal communications,
however. When waiting for the lift, or when chatting during lunch or before
the start of a meeting, you can take these opportunities to fill people in on
progress and make them aware of future events. During the course of your
day, it is likely you will need to make presentations, hold meetings, make
numerous telephone calls, and, generally, have scores of conversations. Do
not view this as a waste of your time. In one sense, it is your primary duty
as a change leader.

In terms of modes of communication, there are many from which to
choose:

1 Interpersonal: a face-to-face conversation on the way to the lift, in periodic (i.e., regularly occurring) meetings, in special (e.g., project kick-off
  meeting) meetings, or informal 'chat' sessions.
2 Non-verbal: individual body language and tone, as well as organisational
  'body language and tone' referred to as symbolic communication
  (e.g., who gets recognition).
3 White papers/reports: a document or report written and circulated to provide information and/or to persuade.
4 Business letter: a traditional, formal written communication mechanism.
5 Memos: Dwyer (1999) outlines five types of memos.
   (a) Instruction memos are intended to provide information needed to carry
       out certain directions (e.g., 'A new photocopier has been installed. All
       staff are welcome to use it. In order to get an access code ... ').
   (b) Request memos ask the receiver or provide certain information or to
       take certain actions.
   (c) Announcement memos provide information.
   (d) Transmittal memos act as a 'cover note' for more formal, lengthy
       messages (e.g., 'The statistics you requested are attached').
   (e) Authorization memos give permission or authority.

6 E-mail, voice-mail: forms of electronic asynchronous communication that I assume are familiar to us all.

7 Newsletters: traditional forms of periodic, newspaper-like, written communications usually sent to the members of a particular organisation, industry or other select group.

8 BTV/video: satellite delivered business television is a relatively new means of communicating within an organisation, often across multiple locations. Business television enables companies to broadcast their own programmes on a private satellite network.

In summary, there are several challenges involved in communicating for change, as illustrated in Figure 4.1.

Figure 4.1 illustrates the premise that in order to get change, represented by the delta symbol (the Greek letter D), we must have a vision of where we want to go, some incentive to get there, the necessary resources and skills, and a plan of action. The lines that follow are intended to illustrate what the likely outcome will be, if a certain element of the equation is missing. For example, if people do not have a clear idea of the desired end-state (vision), there will be confusion. If people lack incentive and skills, there will be no (or slow) change and anxiety. Communication for change, therefore, requires all of this information to be provided, loudly, clearly and regularly. Other information to be communicated has to do with what is not going to change. People's efforts must be paced and rewarded. People need to know what it is they can count on and that they are appreciated. Communication for change is essential for implementation success.

$$\text{Vision} + \text{Incentive} + \text{Resources} + \text{Skills} + \text{Action Plan} = \Delta$$
$$? + \text{Incentive} + \text{Resources} + \text{Skills} + \text{Action Plan} = \text{confusion}$$
$$\text{Vision} + ? + \text{Resources} + \text{Skills} + \text{Action Plan} = \text{no/slow } \Delta$$
$$\text{Vision} + \text{Incentive} + ? + \text{Skills} + \text{Action Plan} = \text{frustration}$$
$$\text{Vision} + \text{Incentive} + \text{Resources} + ? + \text{Action Plan} = \text{anxiety}$$
$$\text{Vision} + \text{Incentive} + \text{Resources} + \text{Skills} + ? = \text{false starts}$$

Figure 4.1   Some elements necessary for change

Communication Strategy

Munter (2000) suggests there are several critical elements to a good communication strategy. A good communication strategy considers all of the following:

1 Communicator strategy. What are your objectives? What communication style best can achieve those objectives? What is the level of credibility of the communicator?

2 Audience strategy. Who is your audience? What do they know and feel? How can you motivate them to action?

3 Message strategy. How should you structure and organise your message(s)? How can you emphasise the key point to increase the likelihood of their being perceived and remembered?

4 Channel choice strategy. Should the communication be in written or verbal form? Passed to a group or one-to-one? Done electronically, via telephone or on paper?

5 Culture strategy. As mentioned near the start of this chapter, every aspect of your communication strategy will be greatly influenced by the cultural context in which you are communicating.

6 Response strategy. While the response is part of Munter's model, she does not address this issue separately. It is important to think about what response you want, if any, from your audience. This can help you focus the rest of the elements of your strategy. For example, if you want people to publicly commit to a certain course of action, you might choose to give a persuasive, emotional presentation at a group meeting so people can have the chance to express their support immediately.

Dealing with the Past and Emotions

In unpredictable and highly contested environments, organisational success requires the ability to implement change more quickly and effectively than your competitors. In order to make changes quickly and effectively people need to be able to let go of the past and grab hold of the future. Most change advice, methods and models, however, focus on ways to help people envision and buy into the future. They rarely, if ever, deal with the past, emotions and the 'grieving' process some people must go through before they can let go and move on.

Change is an emotional process. People fear the unknown, being wrong, unaccepted, ridiculed, embarrassed and, therefore, frequently resist change out of fear. Many people have also become understandably cynical about change. We have seen management tools and techniques come and go (e.g., Quality Circles, Total Quality Management, Business Process

Engineering, down-sizing, out-sourcing, customer relations management) and have come to think, 'This too shall pass.'

I suggest that if we do not deal with this emotional legacy, we will continue to see slow and/or unsuccessful change. In late 2001, I conducted a 'quick-and-dirty' on-line survey of change practices. When I asked change managers which one factor would help them most in their change efforts if they could do it faster, 15 of the 24 respondents (over 60 per cent) answered overcoming resistance and building commitment. I also asked people to estimate the amount of time that was spent on dealing with the 'emotional side' of change (e.g., people's fears or anger over past failed change efforts). On average, people reported spending about 13 days dealing with the 'emotional side' of change in successful projects, compared with 2.5 days in unsuccessful change projects.

Emotional Intelligence at Work

Gardner (1983, 1993) hypothesised that there are actually seven forms of intelligence rather than just one. He suggested that in addition to the traditional logical-mathematical intelligence (usually referred to as IQ, or intelligence quotient), there is linguistic intelligence (the ability to learn and speak many languages), spatial intelligence (the ability to manipulate and/or create mental images), musical intelligence (the ability to write, play and think musically), bodily or kinaesthetic intelligence (awareness of one's body in space and the ability to move), interpersonal intelligence (the ability to know, understand and relate to others) and intrapersonal intelligence (the ability to know and understand oneself).

Goleman (1995) built upon Gardner's work and expanded his sixth and seventh intelligence into what is now popularly known as emotional intelligence. There are six dimensions that characterise emotional intelligence:

- knowing your feelings and using them to make life decisions you can live with
- being able to manage your emotional life without being hijacked by it (i.e., not being paralysed by depression or worry, or swept away by anger)
- being optimistic, and persisting in the face of setbacks and channelling your impulses in order to pursue your goals
- being able to delay gratification in pursuit of goals, instead of getting carried away by impulses
- empathy: understanding other peoples' emotions without their having to tell you what they are feeling

- handling feelings in relationships with skill and harmony (being able to articulate the unspoken pulse of a group, for example)

Let us take a look at a 'light-hearted' example and see how emotional intelligence might show up at work:

> Imagine you are discussing your plans for the implementation of your new IT system with your CIO [Chief Information Officer], or a senior technology executive, who says to you, 'You want me to give you money and take up time we don't have on emotional intelligence training for your people and on helping them deal with their emotions as part of our implementation? Are you crazy or are you just kidding me? This is a business, not group therapy. Besides, emotions have nothing to do with this. This is a technical/business issue.' How would you respond?

Clearly, in this example reacting with anger is not the emotionally intelligent thing to do.

Let us look option a more serious example (Goleman, 1995):

> Assume you have just received the results of your yearly performance appraisal. You had hoped to get a top rating. Instead, you have just found out you got a low-average rating. What do you do?

(a) Sketch out a specific plan for ways to improve your performance and resolve to follow through on your plans.

(b) Resolve to do better in the future.

(c) Tell yourself it really doesn't matter much how you do in your performance appraisal, and concentrate instead on other things.

(d) Go to see the assessor, frankly explain your disappointment, and try to talk him/her into giving you a better rating.

One dimension of emotional intelligence is self-motivation and the ability to be able to formulate a plan for overcoming obstacles and frustrations, and to follow through on it (Goleman, 1995). The best (i.e., most emotionally intelligent) response to the situation above would be option (a).

Consider another scenario (Goleman, 1995):

> Imagine you are in telesales, calling prospective clients. Fifteen people in a row have hung up on you, and you're getting discouraged. What do you do?

(a) Call it a day and hope you have better luck tomorrow.

(b) Assess qualities in yourself that may be undermining your ability to make a sale.

(c) Try something new in the next call, and keep plugging away.

(d) Consider another line of work.

Optimism, a mark of emotional intelligence, leads people to see setbacks as challenges from which they can learn, and to persist, trying out new approaches rather than giving up, blaming themselves, or getting demoralised (Goleman, 1995). While option (a) might seem a good way to go, option (c) would be the most emotionally intelligent answer in the short term and option (b) would be a good longer-term solution.

Research into emotional intelligence suggests that it is an important factor in organisational change and in organisations more generally. For example, research by Huy (1999) suggests that emotional intelligence can facilitate change and social adaptation at the individual level, and that attributes of emotional capability can facilitate radical change at the organisational level. Other research by Goleman, Boyatzis and McKee (2001) and Sosik and Megerian (1999) illustrates that when leaders are more self-aware, the cornerstone of emotional intelligence, they are rated as performing better and rated as exhibiting more transformational leadership behaviours. The same studies showed that more self-aware leaders also perform better in terms of effectiveness, effort and employee satisfaction.

Durscat and Wolff (2001) have taken the concept of emotional intelligence and given it a 'twist' by looking at the emotional intelligence of teams. They defined individual emotional intelligence as (1) self-awareness of emotions and the ability to regulate them, and (2) awareness of the emotions of others. They similarly defined group emotional intelligence as (1) the group's collective self-awareness of the emotions of the individuals in it and, therefore, awareness of the group's emotions overall, and (2) the group's awareness of the emotions of other individuals and other groups. They found that greater individual emotional intelligence leads to greater group emotional intelligence, which in turn leads to higher levels of trust, identity, and efficacy within the group. This leads to greater levels of group participation, cooperation and collaboration which, in turn, leads to better decisions, more creative solutions, and higher group productivity.

Issues for Implementation Success

We know from decades of research on the topic that there are several factors which have been shown to be particularly important to the successful implementation of organisational innovation and technical change. To end this chapter, therefore, we will take a look at a number of issues that will probably affect the success of your implementation efforts.

Education and Training

Moran and Reisenberger (1994), identified some of the competencies necessary for global managers:

- a global mind-set
- the ability to work as an equal with people from diverse backgrounds
- a long-term orientation
- the ability to facilitate organisational change
- the ability to create learning systems
- the ability to motivate employees to excellence

Few of us are born with these competencies. It is essential, therefore, that managers, whether they are involved in significant innovation and technical change or not, continually seek to upgrade their skills and abilities.

In fact, Fulmer, Gibbs and Goldsmith (2000) addressed the question of how companies such as General Electric, Hewlett-Packard and Johnson & Johnson keep a steady stream of leaders moving up. Their data suggest the answer is that people in these companies believe that leaders who keep learning are the ultimate source of sustainable competitive advantage. These organisations also focus on what the authors referred to as the five essentials of leadership development:

1 They are keenly aware of external challenges, emerging business opportunities and strategies, internal developmental needs and the ways other leading organisations handle development.

2 They emphasise the future. Top leadership-development companies use anticipatory learning tools.

3 They recognise that action, not knowledge, is the goal of leadership-development processes. They bring the world into the classroom, applying real-time business issues to skill development. They recognise that answers to tough questions are not in the instructor's head: learners must discover them during the process.

4 They recognise the importance of alignment between leadership development and other corporate functions and they often tie educational efforts to formal succession planning.

5 They seriously assess the effect of leadership programmes on business results.

The defining challenge for competitiveness has shifted from low cost and high quality to the creation and commercialisation of a stream of new products, new services and new processes that shift the technology frontier, progressing as fast as rivals catch up. The drivers of this non-stop innovation, according to Harvard's strategy guru, Michael Porter, and MIT's Scott Stern (2001), are, among other things, a common innovation infrastructure that sets the basic conditions for innovation including a critical mass of people with appropriate levels of education and technical skills.

As stated in Carlopio (1998):

> Frequently, one of the main roadblocks to successful implementation is a lack of understanding and appreciation of the new technology or innovation at all levels in the organisation. Almost everyone who comes in contact with innovation and technological change needs to gain some new skills, new perspectives, new attitudes, and so on. Without new skills, people will be anxious and unable to take full advantage of new technology. Without changes in attitudes and beliefs (i.e. personal change), people will not be able to make the necessary adjustments that make organisational change and the implementation of innovation and technical change happen. The importance of education and training for workplace innovation and technical change cannot be over emphasised.

I suggest that training is so important that it actually needs to happen at three different times during the implementation process. The first round of education and training should be focused on simple awareness-raising. This should happen well before the change takes place. People need to be told the 5 Ws (i.e., who, what, where, how, why, and when; I know there are six and one is an 'H', but we still call them the 5 Ws) in relation to change. For example, people need to know that a new technology is being considered and will be implemented sometime in the future or that a merger or organisational restructure is being contemplated. The second round of education and training should be more focused on operational issues related to the new technology, new strategy or innovation. This should happen close to the time that the change is installed (see Chapter 7). For example, if new technology is being implemented, operational training should focus on the basics of how to use the new technology. If it is a non-technical change, education and training should focus on the operational implications of the change (e.g., the operational implications of a merger and acquisition, or a new business model). The third round of education and training should happen some time after installation, and should be focused on 'advanced' issues. For technology, once people have mastered the basics, they need to know how to use the more sophisticated or advanced elements, features and functions of the system and/or software. For non-technical changes, people need to consider how best to fully take advantage of the innovation, now that they have had time to deal with the basics.

Top-down or Bottom-up?

We know that the higher the organisational level at which managers define a problem or a need, the greater the probability of successful implementation (cf. Bell and Burnham, 1987; Koehler, 1992; Long, 2000; May and Kettelhut, 1996). It is necessary to have senior management sponsorship, commitment and involvement. It is also essential that there be, in a relatively senior position, a project champion. The champion does at least two things: (1) s/he influences the design of the system to ensure the innovation or technical change meets user and business needs, and (2) s/he provides the impetus for implementation of the project (e.g., direction, resources and motivation). At the same time, however, we also know that the closer the definition and solution of problems are to end-users, the greater the probability of successful implementation. This 'paradox' requires consideration and continual management. For years, research has highlighted three major elements that are consistently linked to successful implementation (Bikson and Gutek, 1983; Roman, 1986):

- top management must support the effort, but not define the procedures to be used
- the technicians involved should provide expertise and computer resources, but not be in charge of the implementation
- the 'users' should manage the implementation, but must ensure coordination with both top management and technical personnel

Prototyping and/or pilot projects?

Wherever possible, prototyping/pilot projects should be used. Whether implementing self-directed learning (Piskurich, 1994) or sales force automation (Gondert, 1993), pilots are essential for success. Rapp and McCubbin (1997) suggested that a small, manageable pilot programme which garners and maintains support from senior management is necessary to ensure successful implementation. An analysis of sales force automation failures revealed that one of the ten most common mistakes is bypassing the pilot test (Gondert, 1993).

In the early stages of a project, the use of a small system with restricted scope and/or functionality is frequently associated with success. Pilot projects that are limited in time, scope and resources can be used as training grounds, and as a test bed in which to experiment. When dealing with complex implementation projects, it is not reasonable to expect that people can get it done perfectly the first time. This can help to reduce risks and, once successful, can be used as a showcase.

CHAPTER 5

Segmenting the Internal Market

Our aim in this chapter is to begin to develop sophisticated implementation strategies tailored to the specific innovation we are trying to implement as well as the 'local' conditions within which it will operate. The good news is that we will continue to refine these strategies throughout the remainder of this book. As you should be aware by this time, this is an ongoing process, not a one-off task.

It is important that we leave behind the idea that it is 'best practice' to develop a generic, one-size-fits-all implementation strategy. We must consider the characteristics of the innovation and the individuals involved, as well as the cultural and structural environments within which they operate, so we can develop customised implementation strategies and plans for various groups of users and technologies. We must consider the organisational structure and culture of the organisation and adapt our strategies to these 'local' conditions as well.

In order to enable change to take place, change managers must provide new skills, new structures and sufficient resources to motivate people to behave in new ways. There are so many factors reinforcing the status quo that change-enabling processes (i.e., social-psychological facilitating structures) must be well defined, well funded and well supported. There are a host of issues to consider at this point having to do with rewards and measures, users' perceptions and preferences, as well as organisational structures and culture. Some of these are organisational levers managers must adjust to help get new technology implemented (e.g., measures and rewards, covered in more detail in Chapters 6 and 8). Others are factors that form the core of this chapter, which must be taken into account when developing implementation strategies and tactics.

Studies have shown that different types of people have different capacities for, or interest in, change (cf. Rogers, 1995). Studies have also identified characteristics of innovations and new technologies that affect their likelihood of successfully being adopted (cf. Tornatzky and Klein, 1982; Zaltman, Duncan and Holbek, 1973). Data have also revealed that not every organisation structure and culture is the same as every other in its capacity to cope with, and successfully integrate, change (cf. Burns and Stalker, 1961; Claver, Llopis, Garcia and Molina, 1998; Smith, 1998).

A basic tenet underlying marketing strategy is that there are distinct market segments, each of which has its own needs, desires and interests. The idea is that highlighting aspects of a product or service that are especially appealing to a market segment can increase sales. For example, in the automobile industry, it is recognised that some people value performance while others value safety. When appealing to the safety-conscious, marketers focus on different aspects of their product (e.g., special construction and design features) from those used when they are appealing to those who want performance (e.g., speed and handing). Marketers can try to directly identify people who have these different desires and interests, or they can identify certain demographic characteristics that seem to be closely associated with them (e.g., gender, age, income). They then devise marketing strategies to get the appropriate message to the appropriate group via the most effective channels. They determine if a certain group is more likely to watch television or read the newspaper. If they read the newspaper, they try to determine which paper and which sections of the paper. In order to maximise their chances of success, they then place an advertisement highlighting the features and benefits that are most likely to appeal to that group in the section of the newspaper they are most likely to read.

This concept of market segmentation can be applied to the internal market of potential users of a certain technology within your organisation, as well as to the external consumer market as discussed above. Not all of your users have the same needs, desires or interests, and neither do they all perceive the innovation or technical change you are proposing in the same way.

Your task is to ask yourself the following types of question:

1 How can I discover the distinct, natural groupings (i.e., segments) of people within the organisation?
2 What are the value propositions (i.e., user needs) in each market segment/user grouping?
3 How large are the segments?
4 Who is significantly present/absent in each segment?

Do not over-segment your market. There are no 'hard and fast' rules regarding how many segments are too many. Conceptually, I find it hard to usefully distinguish more than 3–5 internal market segments in most circumstances. For example, in a large international bank, with over 15,000 employees, we found the following segmentation useful:

● head office
● the branch network
● the information technology (IT) group

- the transaction processing centres
- the business banking group

We found different levels of innovativeness, different perceptions of the innovation, as well as distinctly different organisational structure and culture profiles within each of these five groupings.

In a university department, with approximately 350 students, staff and faculty, we restricted the segmentation to:

- faculty
- staff
- students

Again, there were different levels of innovativeness, different perceptions of the innovation, as well as distinctly different organisational structure and culture profiles within each of these three groupings.

Micro-Level Internal Market Segmentation

In this section, therefore, we begin by looking at two ways of segmenting our internal market. Just as organisational strategists and marketers think in terms of market segmentation, so the managers of innovation and technical change must realise that not everyone has the same capacity to change or interest in change. Similarly, not every innovation/change is as likely as any other to be accepted. We will, therefore, also examine several characteristics of innovations and new technologies that have been shown to affect the likelihood of their being adopted.

Innovation Analysis: Stakeholders' Perceptions

There are several characteristics of new technologies that affect the likelihood of their being adopted by individuals and organisations. Research has illustrated that the higher an innovation's score on each of these characteristics, the more successful the implementation is likely to be (Rogers, 1995; Tornatzky and Klein, 1982). One way to segment the internal market, therefore, is based on how different groups of people perceive these characteristics.

The six characteristics are as set out below:

1 Relative advantage – is it better? The degree to which an innovation is perceived as being better than that which precedes it.

2 Compatibility – does it fit? The degree to which an innovation is consistent with existing culture and values, past experience, and current needs.

3 Complexity – can users understand it? The degree to which an innovation is relatively difficult to understand and use.

4 Trialability – can I try it out? The degree to which an innovation may be experimented with on a trial basis.

5 Observability – can I see the operations and results? The degree to which the operations and results of an innovation are observable to others.

6 Re-invention – can I modify it? The degree to which an innovation is changeable or modifiable by the user(s) in the process of its adoption and implementation.

You can use the innovation analysis score card (see Exhibit 5.1 and/or http://www.implementer.com/implementer/web/frames_top.htm) to help compare different subgroups' assessments of various innovation and technical change options. It is essential, if you are the implementer of workplace innovation or technical change, that you do this analysis from the perspective of various 'user' groups, as well as your own. It is the users' perceptions of the characteristics that determine the likelihood of adoption. Because of your involvement with the changes, you will probably have a significantly different perspective of it from that of the eventual users. Different subgroupings within the organisations will, most likely, have different perceptions. It is important that you become aware of these differences if they exist and then tailor your implementation efforts accordingly.

There are at least two ways to do this analysis. The first is to keep a single innovation or technical change in mind, and then identify variations in how people in different subgroups perceive the innovation or changes across the six characteristics. Take the large bank mentioned above, for example. If we are going to implement on-line training for all employees by installing personal computers in every branch and in many offices, we will want to anticipate the reactions of various groups of people so we can tailor our implementation strategy to best suit their needs. We identified that head office and IT saw the relative advantage, thought complexity was moderate to low and that compatibility was high. The branch network and the transaction processing centres, on the other hand, were not clear about the relative advantage (in fact they expect staff cuts as a result), thought complexity was moderate to high and that compatibility was very low. We clearly needed two different implementation strategies requiring significant variations in what and how we would communicate, educate and train, and pilot-test.

The second way to do an innovation analysis is to keep a single subgroup in mind, and then compare their perceptions of two or more different innovations or technologies. For example, if you are going to make two changes, and you want to tackle the most likely to succeed first, you could assess people's perceptions of the six characteristics for each to help you decide which has the best chance of successful adoption.

Exhibit 5.1   Innovation analysis score card

Characteristics (Use a five-point scale ranging from High = 5 through Moderate = 3 to Low = 1)	Innovation name			Innovation name			Innovation name		
	Gp1	Gp2	Gp3	Gp1	Gp2	Gp3	Gp1	Gp2	Gp3
Relative advantage – is it perceived as better?									
Compatibility – does it fit values, past experience, culture?									
Complexity – can users understand it?									
Trialability – can I try it out?									
Observability – can operations/results be seen?									
Re-invention – can I modify it?									
Total score (higher number is better)									

- If relative advantage is low, focus on knowledge and awareness (i.e., communication and training), pilot-test if possible, and align measures and rewards.
- If compatibility is low, focus on adjusting the organisational culture, knowledge and awareness (i.e., processing the past), align measures and rewards, and pilot-test if possible.
- If user understanding is low (i.e., complexity is high), focus on knowledge and awareness (i.e., communication and training), pilot-test if possible, and align measures and rewards.
- If trialability is low, find others who have done it/used it and talk to/visit them.
- If observability is low, focus on measures and rewards, and pilot-test if possible.
- If re-invention is low, focus on knowledge and awareness (i.e., communication and training). If you want re-invention low (i.e., standardisation), explain why. If it is low and you do not want it that way, explain why it cannot be modified.

Innovator Analysis: Stakeholders' Innovativeness

Another way to segment the market is based on how people behave or think. People and organisations adapt to change at different rates. This has led to the idea of individual and organisation innovativeness. Innovativeness is an important concept and has received a great deal of attention recently. For example, Johnson, Donohue, Atkin and Johnson (2001) suggest that organisational innovativeness is becoming the single most important issue in determining ultimate success. Han, Kim and Srivastava (1998) illustrated that an organisation's innovativeness positively influences its business performance and suggest that innovativeness is the most effective means to deal with turbulent, uncertain external environments. Hurley and Hult (1998) found that higher levels of organisational innovativeness (i.e., openness to new ideas) are associated with a greater capacity for adaptation and the ability to successfully implement innovation.

Goldsmith, d'Hauteville and Flynn (1998) and Johnson *et al.* (2001), as well as Rogers (1995), have illustrated that innovativeness is a valid construct at both the individual and organisational level which can be measured reliably. Rogers (1995) identified adopter categories based on the behaviour of people and then discussed their socio-economic characteristics, associated personality variables and communication behaviour. He even discussed 'audience segmentation' and the potential advantages of communicating different benefits of an innovation to different 'sub-audiences'.

This is the method we will use to segment our internal market in terms of innovativeness. After familiarising yourself with each of the categories, you can conduct an innovator analysis to identify natural subgroups of people based on their level of innovativeness. Rogers' (1995) five categories are set out below:

1 Innovators. Innovators are venturesome. Whether they are innovative firms or individuals they have the time and money necessary to innovate. They also have the ability to cope with a high degree of uncertainty and are risk takers.

2 Early adopters. Early adopters are the embodiment of successful and discrete use of new ideas (Rogers, 1995). They act as opinion leaders and are close enough to the mainstream to act as role models. They are predisposed to try something new.

3 Early majority. The early majority are not naturally inclined to try something new simply because it is new. 'They follow with deliberate willingness in adopting innovations, but seldom lead' (Rogers, 1995, p. 265).

4 Late majority. The late majority are sceptical about change. The pressure of peers is often necessary to motivate adoption. Most of the uncertainty

about a new idea must be removed before they feel it is safe to adopt (Rogers, 1995).

5 Laggards. The laggards are resistant to change and change agents. Their point of reference is the past. In other words, 'If we have never done it that way before, why should we do it that way now?'

There are at least two ways to determine which individuals, groups and/or organisations fall into which categories. The first way is to do it 'intuitively'. You, as an individual or with several others, can sort yourself, a group, and/or an organisation into the different categories based on your impressions. While not at all scientific, this method is frequently satisfactory as it is relatively easy to identify the extremes, and that is, in most cases, good enough for our purposes. In other words, if you think about the potential users/adopters, in many cases you 'know' who is routinely positive (i.e., innovators and early adopters) and negative (i.e., the laggards) because of how they have behaved in the past and because you know them. Everyone else, who is not in one of the extreme groups, is 'in the middle' and that is fine. One word of warning: I have done this analysis with hundreds of groups from scores of organisations, and have seen many people fall into the trap of thinking that everyone who is 'young' is innovative and everyone who is 'old' is not. Research by Szmigin and Carrigan (2001) clearly illustrates that there are young innovators and old innovators. Base your classifications of individuals, as much as possible, on how they have routinely reacted to innovation and change in the past.

Another way to sort people into the adopter categories is to use a more formal questionnaire-based method (see http://www.implementer.com/implementer/web/frames_top.htm). While this may seem to be a far superior method to the 'intuitive' method, it is not necessarily so. Just because we attached numbers to the analysis, this does not change the fact that it is still a subjective assessment process. Unless you get very large numbers of people to fill out the questionnaires, this method of assessing the levels of innovativeness in different users (i.e., individuals, groups and subgroups) may not be very much more valid or reliable than the 'intuitive' method.

It is important to remember that this adopter categorisation should always be done along with the previously discussed innovation analysis as various stakeholders' perceptions could account for an innovative individual acting as if he or she were a laggard. In other words, I might be, in terms of my basic personality, a truly innovative individual, but because I do not perceive any relative advantage to a particular piece of technology or software, I might act like a laggard.

Once you have categorised your 'users' you can adapt your implementation strategy so as to capitalise on people's natural tendencies, as explained below:

1 Innovators simply need access. If your are trying to get innovators to adopt a new technology, product or idea, just give them access to it and get out of the way. Involve innovators in pilot tests if you need quick wins, but be aware that they may not be representative of the majority of your population.

2 Early adopters need to know it is the right thing to do before they will adopt. As with the innovators, early adopters are predisposed to change. Early adopters, however, require some information, education, reasoning, examples and involvement to convince them to innovate. If you fail to convince them it will be catastrophic to your efforts because they are close enough to the mainstream to act as role models for those who follow.

3 Early majority. The early majority require even more information, education, reasoning and examples, as well as the use of role models (i.e., early adopters) and involvement to get them to adopt.

4 Late majority. In order to get them to adopt a change, you must reduce their uncertainty and fear by providing a sound financial case and by showing them how the technology works in other parts of the organisation and/or in other organisations. Education and training, pilot-testing and prototyping are important to convince them to make changes.

5 Laggards. The laggards require maximal information, education, reasoning, examples, pilot-testing and prototyping, role models and involvement to get them to adopt. Even with all your efforts, you still may not get all laggards to adopt.

Macro-Level Internal Market Segmentation

The first of our two macro, organisational-level variables that can be used for internal market segmentation is organisation culture. Organisational culture is considered by many people to be a 'soft' or a 'fuzzy' topic. It is hard for some people to get a good grasp on what it is. This is unfortunate, as research has consistently linked elements of organisational culture to successful economic performance (cf. Bolman and Deal, 1997; Collins and Porras, 1998; Denison and Mishra, 1995; Han, Kim and Srivastava, 1998; Higgins, 1995; Kotter and Heskett, 1992; Smith, 1998).

The second organisational-level variable that can be used for internal market segmentation is organisational structure. Organisational structures vary in their capacity to implement or generate innovation and technical change. This relationship between structure and technology is not one-way, however: that is, not only can certain organisational forms affect the rates and success of innovation and technical changes, but new technology can have an impact on the existing organisational structures in adopting firms.

Organisational Culture

There are three broad conceptualisations of organisation culture with which you should be familiar. The external perspective views culture as a background factor influencing people's values, beliefs and pre-dispositions about organisations, work and authority. Organisation culture is developed and maintained outside organisations and is akin to what many think of as national culture. The internal perspective views culture as the set of unique rituals, legends and ceremonies developed within an organisation and influenced by its structure, size, technology and leadership. Finally, the root metaphor perspective suggests that an organisation does not have a culture: an organisation *is* a culture. Each of these will be briefly discussed in turn.

The external perspective
From this perspective, organisational behaviour, processes, structures and systems are influenced by prevailing national culture. Recent work (cf. Trompenaars and Hampden-Turner, 1998) suggests that national culture affects people's perceptions and expectations and is a major determinant of how people decide to resolve the many dilemmas facing organisations and managers. The issue of cultural diversity in organisations has received much attention recently and is based on this conceptualisation of organisational culture. The concepts of cultural relativism (i.e., *the* correct way to perceive, think and feel; the right way, the answer, manners, what you should and have to do) and ethnocentrism (i.e., the tendency to regard one's own culture as superior to others) stem from work in this area.

I once heard someone say that 'Eastern' and 'Western' management practices are 95 per cent the same, and differ in all important respects. This external view of organisational culture suggests that we can no longer say that the way large North American, British and Australian organisations do something, is right for every organisation on the planet. For example, empowerment and performance-related pay may motivate people in the USA and Australia, as long as they have control over their work inputs and outputs, but why does appealing to the importance of the relationship sometimes work better in Japan, Singapore and China? Trompenaars and Hampden-Turner (1998) suggest this is because of basic cultural differences regarding individualism and collectivism.

Finally, people with this external view of organisational culture are working towards the development of principles of international and transnational management that can be applied across national culture. Trompenaars and Hampden-Turner (1998), for example, suggest that the matrix organisation structure reconciles (in the USA, UK and Scandinavia) the universal dilemma regarding the need for a focus on both the individual discipline or functional area and projects or customer needs. They suggest, however, that the matrix structure threatens and contradicts the family cultural model in Italy, Spain, France and some Asian countries, so those working in these countries need to devise a different structural solution.

The internal perspective
Within the internal perspective on organisational culture, many people view culture in terms of behaviour. The following aspects should be considered:

1 Culture is learned by the connection between behaviour and consequences. People do what they do because of the rewards and 'punishments' they receive.

2 Culture is transmitted through a pattern of behavioural interactions. You can only 'see' culture as it is reflected in what people do, and do not, do.

3 You can assess a culture by asking behaviourally-oriented questions such as:

   (a) How are employees treated?

   (b) How do you want to be seen by outsiders?

   (c) What behaviours are allowed/encouraged?

   (d) Does hard work pay off?

   (e) Does anyone care if you are late?

   (f) Is it OK to express emotion?

4 In an organisational setting there are multiple reinforcements and reinforcing agents. We receive many messages from many sources regarding what is, and is not, acceptable behaviour.

5 Each individual carries pre-dispositions that shape their interpretation of the culture. This, in a sense, acknowledges the existence of the internal perspective discussed above.

6 A symbiotic relationship exists between reinforcing agent and target. In other words, culture is maintained by a self-fulfilling cyclic system.

I behave in a certain way because it is the cultural norm, which further strengthens and communicates the cultural norms.

7 Changing an established culture, therefore, is difficult. Because of all of the above, this is not a simple process to 'short-circuit' or change.

The root metaphor perspective

The root metaphor perspective is concerned with organisational culture and:

- symbol (e.g., size of office signalling who has power)
- myths (e.g., support claims of distinctiveness and foster internal cohesion)
- stories and fairy tales (e.g., provide comfort, reassurance, direction and hope)
- ritual (e.g., routine, day-to-day patterns and sacred habits provide stability)
- ceremony (e.g., special occasions to celebrate and reward)
- metaphor, humour and play (e.g., help us make sense of the unfamiliar, reduce tension, be creative and flexible)

Organisational culture, from this perspective, has also been conceived of as theatre:

1 Organisational structure and hierarchy can be seen as a stage design portraying power relationships, values and myths.

2 Organisational processes – all the world's a stage: meetings, planning and evaluation processes. We 'make believe' that they produce the desired outcomes, but they very often do not. Everyone knows many meetings are a waste of time, the planning process is a 'joke', but we continue to engage in them year after year. This is because they are cultural symbols, games, excuses for interaction, and advertisements.

The importance of organisational culture

Regardless of how you conceptualise organisational culture it is important for many reasons:

1 It has been linked to economic performance and organisational survival (Bolman and Deal, 1997; Collins and Porras, 1998; Denison and Mishra, 1995; Han, Kim and Srivastava, 1998; Higgins, 1995; Kotter and Heskett, 1992; Smith, 1998).

2 How someone becomes a member of the culture is important. A disproportionally large amount of employee turnover occurs early in the period of association. Improving the quality of the early job experience leads to reduce employee turnover (Mowday, Porter and Steers, 1982).

3 Diversity can provide competitive advantage (Cope and Kalantzis, 1997; Robinson and Dechant, 1997).

4 Example, not command, holds the culture together (Bolman and Deal, 1997).

5 Specialised language fosters cohesion and commitment (Bolman and Deal, 1997).

6 Stories carry history and communicate values (Bolman and Deal, 1997).

7 Stories reinforce group identity (Bolman and Deal, 1997).

8 Humour and play reduce tension and encourage creativity (Bolman and Deal, 1997).

9 Ritual and ceremony lift spirits, reinforce values and provide stability in a changing world (Bolman and Deal, 1997).

10 Informal cultural players make contributions disproportionate to their formal roles (Bolman and Deal, 1997). There are key individuals who play pivotal roles (e.g., opinion leaders, change champions, rumour starters) in the organisation, who may not be 'formally' in positions of power.

11 Culture is the 'secret' of success. Culture is the 'magic' ingredient making some organisations especially successful and making some workplaces especially great places to work.

Organisational culture analysis

Organisational culture is the collection of norms, values, beliefs, expectations, assumptions and philosophy of the people within it. Different organisations, and different parts of the same organisation, use diverse jargon, participate in various rituals, and use a number of different artefacts. For example, merchant bankers, as compared to medical practitioners or school teachers, would have a jargon all their own and would use certain analysis tools and techniques vastly dissimilar to the language and tools used by doctors or teachers. Even within the same organisation, there would be distinct subcultures. We would expect people in an engineering or IT department to have a distinctly different culture from those in the accounting or human resources areas.

There are many ways to visualise the concept of organisational culture. One popular conceptualisation is the onion model. If you cut an onion in half and look at it, you will see many layers. An organisation's culture can be visually represented in this way (as illustrated in Figure 5.1).

When we walk around an organisation, there are elements of the organisation's culture that are 'on the surface' and are relatively easily visible. We

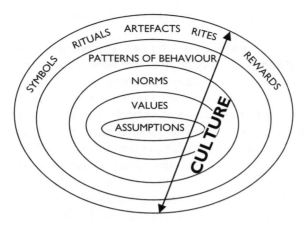

Figure 5.1   The onion model of organisation culture

Figure 5.2

can see many cultural symbols (e.g., whether your office is on a floor close to the top or the bottom of the building, how big your office is), artefacts (e.g., computers), and patterns of behaviour (e.g., how and where people interact, how they behave in formal and informal meetings). Less visible, but equally important, are the less tangible aspects of culture such as the norms, values and basic assumptions people make.

Another way of conceptualising organisational culture is in terms of its 'hard' and its 'soft' sides. As we see in Figure 5.2, organisational culture is 'supported' by both social/psychological aspects (e.g., stories, symbols, rituals) and by some more concrete elements such as power structures, hierarchical structure and control systems (e.g., financial, measurement and reward systems).

While these models provide ways for us to explore many aspects of organisational culture informally, they do so in an unstructured and non-prescriptive manner. For those of us specifically concerned with the strategic implementation of technology, it is important that we focus on the dimensions of organisational culture that are most critical to innovation and successful organisational change (you can find these cultural analysis tools on-line at http://www.implementer.com/implementer/web/frames_top.htm).

From the literature on organisational culture, I have identified four cultural dimensions that have been shown to affect the ability of an organisation to innovate and change: external expansion and search; long-term, future orientation; adaptability; and values orientation. Each of these is outlined below and is presented in a way that allows us to make some prescriptive statements about their impact on the implementation of innovation and technical change.

External expansion and search   Adaptive cultures reach out and search for new information, markets and technology. There are two aspects of external expansion and search that are relevant for the strategic implementation of technology:

> Market expansion – you are always trying to expand into new markets, invest in new technology and emphasise the importance of long-term viability to enhance versatility (Christensen, 1995; Claver *et al.*, 1998; Higgins, 1995; Morris, and Trotter, 1990; Smith, 1998).

> Pro-active information search – you are always looking for information regarding new markets, technologies, and products/services (Easterby-Smith, 1990; Higgins, 1995; Morris and Trotter, 1990).

Does this describe your organisation?

> If yes: people will know what is going on in the marketplace and, therefore, will be more likely to appreciate the need for the strategic innovation you are implementing. This will greatly facilitate your efforts.

> If no: people will probably not know what is going on in the marketplace and, therefore, will be less likely to appreciate the need for what you are suggesting. They will need to be given this information.

Your strategy to neutralise this road block is to focus on communication and education. You must explain what is going on in the market, why, and how your innovation will address this 'need'. You must take the place of the culture that would naturally lead people to search pro-actively for information regarding the market, new technology and new products and services.

Long-term, future orientation   Focusing on the longer term allows firms to invest and take risks reflected in a future-orientation with long-term goals regarding investment and pay-back. There are two important aspects of orientation relevant to the strategic implementation of technology:

> Long-term focus – you assume there will always be increased competition, and you therefore need to have a long-term strategy for expansion and growth, as well as recognising that sometimes a short-term erosion of profit is normal as we invest in the future (Claver *et al.*, 1998; Higgins, 1995).

> Future orientation – you have a predisposition towards constant learning, you value training and development activities and the evolving skill-sets of staff (Claver *et al.*, 1998; Collins and Porras, 1998; Kotter and Heskett, 1992).

Does this describe your organisation?

> If yes: people will be oriented towards the future, will view their relationship with the firm as long term, and will have goals and objectives that promote long-term investments in technology and skills. This will greatly facilitate your efforts.

> If no: people will be oriented towards the short-term, will view their relationship with the firm as short term, and will have goals and objectives that promote short-term thinking, maintenance of the status quo, and reliance on cost controls. Your strategy to neutralise this road block should focus on affecting reward systems and performance indicators, because people notice what you measure and you get what you reward.

Adaptability   Cultures adapt and change to meet evolving and emerging market demands. There are two important aspects of adaptability:

> Risk-taking – you are always looking to develop new products/services and to use new technology in development and production/delivery of your products/services. It is 'OK' to take risks, to experiment, and to learn from failures (Barnes, 1993; Claver *et al.*, 1998; Collins and Porras, 1998; Smith, 1998).

> Flexibility – you actively change to better fit with your external environments. You engage in self-betterment and risk-taking in response to intensifying competition. Change is accepted as the norm (Bahrami, 1992; Claver *et al.*, 1998; Denison and Mishra, 1995; Smith, 1998).

Does this describe your organisation?

> If yes: people will be relatively comfortable with risk. They will be used to continuous change. This will greatly facilitate your implementation efforts.

If no: people will be uncomfortable with risk. They will be more likely to resist change. Your strategy to neutralise this road block should focus on education and training and 'processing the past'. You will need to provide substantial opportunities for people to gain knowledge and skills, and for gaining relevant information. With more knowledge and information, people will become more comfortable with calculated risk-taking. In terms of 'processing the past', people need to know why it is going to be different this time. Why should they trust you? People need to have mechanisms for processing their fears as change is an emotional process. Fear is the greatest obstacle to honesty. People 'dig their heels in' and resist out of fear. Unresolved anger blocks listening and leads to passive-aggression and resistance. This inhibits creativity, spontaneity, energy, expression and communication. It becomes not safe to explore, to challenge, to change, to risk, or to be involved.

Values orientation   It is true that technical innovation will not happen successfully without technical competence. While getting the technical issues correct is a necessary condition for success it is, unfortunately, not sufficient. Implementation is more than just 'plugging it in'. Successful innovation requires cultural and structural changes in the organisation's social system and in individuals. There are two important aspects of values orientation:

Core values – there is a set of shared, core values revolving around participation, empowerment and shared responsibility for innovation (Claver *et al.*, 1998; Collins and Porras, 1998; Higgins, 1995; Kotter and Heskett, 1992; Umiker, 1999).

Focus on creativity – creativity is stimulated and the focus is on the quality of the idea, not the power and authority of the person proposing it (Claver *et al.*, 1998; Higgins, 1995; Umiker, 1999).

Does this describe your organisation?

If yes: people will have the power and will share many of the values necessary to facilitate innovation and change. This will greatly facilitate your efforts.

If no: people will be focused more on position and power than on ideas, innovation and creativity. Your strategy to neutralise this road block should focus on organisational power and structure. It is likely that your organisation will be more hierarchical and mechanistic than flat and organic (see the organisational structure analysis for more on this). If it is possible, you should decentralise the organisational structure and the decision-making power and authority in your organisation. In most cases, however, this will not happen quickly or it will not be possible at all. In the short term, therefore, you should focus on securing resources and on fulfilling important roles. With powerful sponsors and champions, as well as the necessary resources (e.g., money, personnel, facilities, equipment), you can gain most of the power and authority you will need to implement your innovation or changes successfully.

Organisational Structure

Organisation structure has been consistently related to frequency and ease of technological innovation (cf. Burns and Stalker, 1961; Claver *et al.*, 1998; Perry, 1995). Firms having organic structural forms tend to be more technologically innovative. Some specific characteristics of organic forms of organisation that are frequently related to innovation are: (1) informal definitions of jobs; (2) lateral, network-like, communication patterns; (3) consultative rather than authoritative communication styles; (4) diffusion of knowledge-seeking throughout the firm; (5) greater prestige and importance attached to extra-organisational knowledge or activities than to internal knowledge or activities.

Firms with more mechanistic structures are usually less technologically innovative and more resistant to change. Some specific characteristics of mechanistic organisational forms are: (1) rigid breakdowns of roles and jobs into functional specialisations; (2) precise definitions of duties, responsibilities and power; (3) hierarchical control, authority, and communication; (4) belief that the managers at the top know what is best; (5) reliance on vertical interactions; (6) greater prestige and importance attached to internal than to extra-organisational knowledge or activities.

The structure of an organisation does not operate independently of its managers. Structure shapes the behaviour of managers and, in turn, is shaped by their behaviour. For example, organisations that are mechanistic select and reward managers with values and behaviours consistent with this form of organisation. These managers then perpetuate the system, while innovative managers never get ahead or choose to leave the firm.

Organisational size frequently shows up as a strong predictor of innovativeness. Even though size is consistently and powerfully linked to innovativeness, this does not tell us very much. Research has shown that size is a surrogate for several other variables such as total resources, slack resources and organisational structure. Consequently, most current research is concentrated on examining more specific structural features of organisations that frequently co-vary with organisational size and are related to technological innovation in complex ways, such as the following:

1 Centralisation of power and decision-making. The more power is concentrated in a few top leaders, the less innovative the organisation is likely to be. However, low centralisation makes it more difficult to implement innovations.

2 Organisational complexity. The higher the levels of employee knowledge and expertise in a firm, and the greater the range of occupational specialties

in a firm, the more innovative the individuals and the organisations are likely to be. However, with higher levels of organisational complexity comes greater difficulty to achieve consensus about implementation.

3 Organisational formalisation. The more rules and regulations are used to govern organisational behaviour, the less likely an organisation is to consider innovating. However, with higher levels of organisational formalisation comes greater ease in implementation. For example, in the military (which is the classic example of a highly formalised organisational structure although they are not on the cutting edge of social innovation), when they do decide to make a change, they simply formalise the change and it is very quickly and totally implemented.

4 Organisational interconnectedness. The more social systems and organisational units are linked by interpersonal networks and relationships, the better the flow of information and ideas, and the more innovative the firm is likely to be.

5 Organisational slack. The more uncommitted resources that are available within an organisation, the more innovative the organisation is likely to be.

Please remember that, with all of the variables discussed above, research does not indicate that they cause a firm to be more innovative. These variables have been shown to inhibit or facilitate innovation and technical change, but not to cause it. In other words, just because a firm has a great deal of extra resources it does not mean that it will necessarily be more innovative than a firm with fewer. Research consistently shows, however, that having more slack resources makes it easier for firms that are trying to innovate to do so successfully. Similarly, the fact that a firm is very low on formalisation does not mean that it will necessarily be innovative. The data show that it is easier to innovate in a firm that is less formalised than it is to innovate in a highly formalised firm.

Organisational structure analysis: mechanistic versus organic structures

The prevailing organisational structure will affect people's ability and willingness to change. This first analysis looks at how mechanistic or organic the organisational structure is overall, and how this will affect your ability to innovate. With this analysis, we will attempt to gain an understanding of how the organisational structure is likely to impact on the success of your implementation.

For each of the structural characteristics listed in Exhibit 5.2 on the left, think of the innovating unit (e.g., group, department, division and organisation) and choose one of the three responses that best describe the unit. This is only a rough analysis so we have limited you to one of three broad choices. Choose response 1 if the unit is best described by the item on the left. Choose response 3 if the unit is better described by the descriptor on the right. Choose option 2 if both descriptors describe the unit equally well or if the characteristic does not apply. In other words, when you consider roles (the first characteristic below), if roles in your organisation are more specialised than broad, you should choose 1. If roles are generally more broad than specialised, then choose 3. If some roles are specialised and some are broad, choose the middle response 2.

If the Grand Total score is between 10 and 16, the organisation should have the following structural attributes.

The structure of the innovating unit is more mechanistic than organic. Research indicates that firms with more mechanistic structures are usually less technologically innovative and more resistant to change. These organisations are known to have rigid and inflexible structures and to be especially unable to cope with changes and dynamic external environments. They are subject to failures of information and communication resulting from over-specialisation and separation of functions that make the transfer and feedback of information difficult. Tasks and relationships are carefully prescribed and impersonal. These structures provide few conditions for individual motivation and self-development.

Mechanistic organisations that find themselves with the need to innovate and to implement significant technical changes must make the transition to new organisational structures, preferably before they implement the changes. Mechanistic organisations that have already begun implementing significant changes with little or no change to their organisational structures will tend to find that the innovations are not working very well.

Either way, you will have to begin to make significant modifications to your organisational structures by broadening roles and freeing up communication and the flow of information. You must focus your attention on facilitating structures (especially change agents, communication, reward systems, roles, resources, and working groups) and on persuasion/decision/commitment. You may also want to consider the following.

If you have implemented new technology without making significant changes to your organisational structure first, you may find that the effects of new technology on the structure and processes of your organisation may take a while to percolate through. At first, new technology may have immediate effects on the contents of jobs and on the skills and qualifications that are necessary to work with it. Over time, the old organisation design becomes suboptimal with respect to the possibilities offered by the new technology.

Exhibit 5.2

Structural characteristics	Mechanistic	Structures: Circle the appropriate score			Organic
		1	2	3	
Roles	Specialised	1	2	3	Broad
Role Definition	Others determine what you will do	1	2	3	What you do is determined by self/peers
Understanding of tasks	Your supervisor knows why you do what you do	1	2	3	Self/peers know why you do what you do
Commitment	To exact performance of duties	1	2	3	To the broad enterprise as a whole
Communications supervisor	Vertical – only to and from	1	2	3	Horizontal – anyone can talk to anyone
Knowledge	Concentrated at top	1	2	3	Diffused throughout
Information	Concentrated at top	1	2	3	Diffused throughout
Interactions	Command and control	1	2	3	Consultation and discussion
Ethos	Organisational loyalty, no public criticism or discussion	1	2	3	Loyalty to the task, technology and/or profession
Orientation	Employee and organisational focus is local	1	2	3	Employee and organisational focus is global
	Totals (sum the columns)				
	Grand total (sum the totals)				

Organisations that successfully manage innovation and technical change will increasingly be concerned with open processes, as well as with rules and procedures. We can no longer solely search for stable rules and procedures and ways to restructure uncertain and unpredictable takes into routines and programmes. We must look at operations as open systems and then maintain them as open processes wherever possible. We must stimulate diversity, interconnection and change, and then value design skills, creativity, quality and variety as sources of added value.

Organisations that can continually innovate and successfully implement change are designed to manage complexity more than to reduce it. In these situations learning becomes critical. We can no longer reduce, subdivide and standardise. We must accept complexity, ambiguity and change and learn to reward them and to use them to our organisational and individual advantage.

Organisation designers must allow for diversity and they must build in flexibility. That is, organisations must reward quality, service, and variety in products, individuals, and processes rather than the blind pursuit of uniformity and increased production in terms of numbers of units.

If the Grand Total score is between 17 and 23, the organisation should have the following structural attributes.

The structure of the innovating unit is relatively balanced between organic and mechanistic characteristics. Given this hybrid configuration, we cannot make a definitive statement in relation to these variables and their likely impact on implementation.

If the Grand Total score is between 24 and 30, the organisation should have the following structural attributes.

The structure of the innovating unit is more organic than mechanistic. Research indicates that firms with organic structures tend to be more innovative. Your organisational structure is not likely to present many significant barriers to the successful adoption of workplace innovation and technical change.

Implementation implications   If your structure is mechanistic (i.e., your score was between 10 and 16), and is in fit with its environment, it does not need to be changed. We know that within mechanistic structures (as compared to more organic structures), commitment is to the exact performance of duties, job roles are more specialised, role definitions are determined by superiors, communications are more vertical, and knowledge and information are more concentrated at 'the top'.

If you decide to implement a strategic or technical change within a mechanistic organisational structure, you need to focus more (1) on changing role definitions and tasks, (2) on communicating the changes 'top down', and (3) on performance measures and reward systems to motivate the changes.

However, if your structure is relatively organic (i.e., your score was between 24 and 30), and is in fit with its environment, it does not need to be changed. If you decide to implement a strategic or technical change within an organic structure, you need to focus more on verbal persuasion, together with 'user' involvement and participation in decision-making and planning.

CHAPTER 6

Persuasion, Decision, Commitment

Much of our focus in the previous two chapters was on ways to communicate to, and segment, our internal market so we could better persuade the people inside our organisation of the benefits of our proposed innovation or technical changes. In this chapter, we focus on the persuasion process more directly.

As was done at the strategic and organisational level (see Part I, Chapters 1–3), people at the local level need to sort through the available information and options, and they need to decide that what is proposed should actually happen (see Part II, Chapters 4–9). If this is done well, people can commit themselves to help make the changes happen. This is not a particular point in time and neither is it anything that can finally be achieved and then set aside.

Our main goal is to reach the point where the individuals and groups affected actually decide to take on the innovation or technical change we are trying to implement. What we are involved in is a persuasion process that began with adequate knowledge and awareness (Chapter 4), was supported and stimulated by numerous facilitating structures (Chapter 5), and flows through to where we finally reach the stage of individual and group decision and commitment. Please remember that although we treat these elements of the process as if they were discrete and linear, they are not; this is a chaotic and iterative process.

In this chapter, we look at three areas. First, we consider commitment, compliance and resistance, and how we can influence or persuade people to overcome resistance and gain at least compliance, if not commitment. Next we consider the related issues of stress, pacing and celebration. It is important to consider the timing of changes, as change is a stressful process. Our ability to pace change, and to celebrate and reward success, will significantly impact on our implementation success. Finally, we will consider some of the most powerful 'persuaders' available to us in organisations: reward and recognition systems. In many ways, this is an integrative chapter. You should find that many of the issues we are discussing at this point are overlapping and interlinked.

Resistance, Compliance, Commitment

Commitment to Change

Ideally, everyone will be committed to the changes we want to make and will do everything in their power to make change happen quickly, smoothly and successfully. Unfortunately, that ideal is virtually never realised. Some people will be committed, some will comply and some will resist change. Nonetheless, striving to gain as broad a base of support and commitment to our proposed changes as possible is our goal.

Commitment is critical to successful implementation. Commitment to a goal, for example, is an important factor contributing to successful achievement of a goal in a variety of circumstances. When people are committed to some outcome, goal or cause, they exhibit greater determination to achieve the goal and a greater willingness to put forth effort to attain that goal (cf. Hollensbe and Guthrie, 2000; Renn, Danehower, Swiercz and Icenogle, 1999; Wetzel, O'Toole and Peterson, 1999). Committed employees are proactive, highly involved, independent contributors with initiative and a well-developed sense of responsibility (Campbell, 2000).

One way to increase commitment, especially in organisations operating in a predominantly 'Western' cultural orientation, is to employ self-managed teams. Kirkman and Shapiro (2001) found that self-managing work teams whose members assign jobs, plan and schedule work, make production- and/or service-related decisions and take action on problems have been linked to increased job satisfaction, commitment and performance.

Work by Chase and Dasu (2001) illustrates that another way to build commitment is through choice. People are happier and more comfortable when they believe they have some control over the process, particularly an uncomfortable one. For example, blood donors perceived significantly less discomfort when they were allowed to select the arm from which the blood would be drawn. Giving people choice, even when the choice is largely symbolic, as with the blood donors, reduces people's feelings of helplessness and hopelessness, and increases their commitment to make the process work. Chase and Dasu (2001) also suggested that another potential way to increase commitment is to give people rituals and stick to them. People find comfort, order and meaning in repetitive, familiar activities (cf. Zajonc, 2001). We are creatures of habit and we notice deviation from the routine.

Pierce, Kostova and Dirks (2001) explored the major mechanisms through which psychological ownership, or commitment, emerges in organisations. As with the work by Chase and Dasu (2001), perceptions of control appear to be a key characteristic. Jobs that provide greater autonomy, for example, allow people higher levels of control and increase the likelihood that feelings of ownership will emerge, while centralisation of decision-making tends to minimise the amount of control an individual can hold and decreases psychological ownership. With greater commitment and psychological ownership come increased stewardship, citizenship behaviours, personal sacrifice and the assumption of risk. Finally, their study suggests that individuals will be more likely to promote change and feel ownership of the change, if the change is self-initiated, and they will resist change when the change is imposed because it is seen as threatening to their sense of control.

A study by Ruppel and Harrington (2000) illustrated that the greater the level of trust in the organisational subunit (e.g., department) the greater the commitment and innovativeness of the subunit. This finding supports the idea that increasing trust will heighten employees' commitment and willingness to take risks which, in turn, lead to greater creativity and innovativeness. According to Ruppel and Harrington (2000, p. 324):

Less monitoring and defensive behavior by managers and more employee enthusiasm for innovation are believed to be the underlying mechanisms by which trust influences innovation. Thus, this study suggests that the establishment of trust within the organisation is a worthwhile effort in organisations where innovation is desired. Moreover, such effort benefits not only primary stakeholders (e.g., stockholders), but also extended stakeholders (e.g., employees).

The study illustrated that the more open communication among managers and employees was, the greater the level of trust, and the more a corporate climate emphasises human relations and employee interests, the greater the openness of employee communications which lead to more trust and greater commitment and innovativeness.

Resistance to Change

The 'flip-side' of commitment to change is, of course, resistance to it (Coetsee, 1999). It is important to remember that, under most circumstances, some level of resistance is normal and functional. As mentioned above, we are creatures of habit. When something becomes habit, it requires less effort (both cognitive and physical) on our part and is naturally resistant to change.

Resistance to change is a complex, multidimensional response with emotional, cognitive, and intentional components (Piderit, 2000). For example, because of the change process (e.g., no participation in the process or not enough communication), we may be emotionally resistant to the changes, even though we think (cognitive) the changes make good business sense. Alternatively, we may be initially enthusiastic (emotional) and clearly see the need for change (cognitive), but because we are not given the support we expect and think we need in order to make the changes happen, we give up during implementation (i.e., our intentions change). It is rare that employees are all negative or all positive across the three dimensions. It is important to remember that resistance to change is normal and frequently functional. According to Piderit (2000), 'Moving too quickly toward congruent positive attitudes toward a proposed change might cut off the discussion and improvisation that may be necessary for revising the initial change proposal in an adaptive manner.' In other words, discussion, disagreement and experimentation consistently lead to more successful change and are not necessarily signs of resistance.

One of the stereotypical sources of resistance to change in organisations is middle management. Huy (2001a) examined the questions:

> Are middle managers really people who stubbornly defend the status quo because they are too unimaginative to dream up anything better? Do they really sabotage attempts to change for the better or are they corner-office executives' most effective allies when it's time to make a major change in a business?

According to Huy's six-year study of over 200 mid-level managers in organisations undergoing major change, the stereotypes of change-resistant middle managers are largely unfounded. Huy found that middle managers often have value-adding entrepreneurial ideas that they are willing and able to contribute and realise, if only they are given a chance. He also found that they are far better at leveraging the informal networks that make substantive, lasting change possible than are most senior executives. Middle managers are closer to the coalface and, therefore, often stay more attuned to employees' moods and emotional needs than senior executives. This helps ensure that the change initiative's momentum is maintained. Finally, Huy found that middle managers help manage the tension between continuity and change, keeping the organisation from falling into extreme inertia on the one hand or chaos on the other.

In today's global organisations and multicultural, diverse workplaces it is important for us to remember that resistance to change could be due to conflict with values and beliefs that vary across national cultures. A study by Kirkman and Shapiro (2001) illustrated the importance of cultural values when implementing Western-based management initiatives in non-Western affiliates. Their data suggest that employees resist management initiatives when they clash with cultural values. A consistent body of literature has identified differences in levels of resistance to self-managing work teams, for example, as well as differences in satisfaction and commitment across national cultures. Participative management may not be suitable in countries characterised by a belief in the importance of status and power differences (e.g., the Philippines or Russia). Their data also illustrate that in places such as the UK, Australia and the USA, where there is lower pressure for conformity, more freedom to question superiors, and a strong belief that one can take action to effect change, there may be more of a general tendency to resist management initiatives. The authors conclude that 'Attention to, and respect for, differences in cultural values remains a high priority for international managers.'

Another insight regarding resistance to change comes from work done by Harrington, Conner and Horney (2000). Their point is that when problems and opportunities are current (i.e., they are currently in the minds of those involved), change is easier to motivate than when problems and opportunities are anticipated in the future. In other words, if we are in trouble now, there is likely to be less resistance to changes designed to get us out of our current troubles. If we are going to be in trouble in the future (i.e., unless we do X or Y), there is a greater likelihood of more resistance to the proposed changes. In this latter case, more communication, vision and motivation must come from change managers and executives to persuade people of the need for the proposed changes and of the likelihood of the future problems/opportunities. Harrington, Conner and Horney (2000) illustrate this point in a matrix similar to that shown in Figure 6.1.

	Problem	Opportunity
Current	Situation: 'We are in trouble now.' Motivation: The immediate loss of our market dominance, job security, organisational survival, etc.	Situation: 'If we act immediately, we can take advantage of this situation.' Motivation: The loss of a potential advantage that is within our grasp.
Anticipated	Situation: 'We are going to be in trouble.' Motivation: The impending loss of our market dominance, job security, organisational survival, etc.	Situation: 'In the future, we could be in a position to profit from what is going to happen.' Motivation: The loss of a potential advantage that is possible to achieve in the future.

Figure 6.1   The situational impact on resistance

Source: Adapted from Harrington, Conner and Horney (2000)

Harrington, Conner and Horney (2000) also classify the way people express their resistance to change into two categories:

- covert resistance is a marked, concealed reaction to the change
- overt resistance is the expression of the open and honest opposition to the change

They also classify individuals' resistance to change into the following six categories:

1 Arbitrary resistance. This type of resistance comes from people who are against everything. It does not matter if change it is good or bad: they are against it.

2 Justified resistance. This type of resistance comes from people who have realised that a change is going to harm or disadvantage them in some way. As a result, these people do everything possible to prevent it. For example, if a change might cause job losses, those whose jobs are at risk may resist the changes.

3 Informed resistance. In this case, the individuals understand that the change that is going to negatively impact them in some way and they have an idea that will lessen that impact or will make the change more effective. These individuals are not against the change, but they have a strong feeling that the proposed change should be altered to improve the benefits to the organisation's stakeholders.

4 Mistaken resistance. This type of resistance comes from people who are reacting to gossip or false information and, as a result, have turned against the change. These people may change their opinion when provided with the correct information.

5 Uninformed resistance. This type of resistance comes from people who have not been provided enough information about the change. Many people resist changes that they do not understand.

6 Fearful resistance. This type of resistance comes from people who can imagine a number of negative consequences of a change. They are afraid of what might happen to them if there is any change in their familiar environment.

Another source of resistance to change is frequently the over-load and organisational chaos resulting from continuous large-scale change. While it is true that organisations must change to survive, continual change can be so disruptive it can tear an organisation apart (Abrahamson, 2000;

Beer and Nohria, 2000; Huy, 2001b; Wood, 1998). We will further explore this notion of the pace of change in the next section of this chapter.

Stress, Pacing and Celebration

As mentioned at the end of the previous subsection, while change is seen as necessary in most organisations today, too much change can cause resistance, feelings of being overwhelmed, stress and burn out (cf. Abrahamson, 2000; Wood, 1998).

According to Rosabeth Moss Kanter (interviewed in Wood, 1998, p. 92), a professor at the Harvard Business School and author of numerous books including *When Giants Learn to Dance* and *The Change Masters*:

> We need more stability in the workplace. We can't have executives managing by revolutionary change where things are stable for a while and then all of a sudden there is a huge upheaval. This cycle of storms followed by calms is very disturbing to the workforce.
>
> My idea of change is more of a steady, ongoing, yet calm kind of change that is a very important part of a company's success. The goal is to continuously experiment with new products and ideas so it is part of the normal excitement of life in that company. People don't usually think of this as change because it isn't abrupt, scary or linked to a specific event. In this kind of environment people can think about the future and invest their energy in new projects, but they can do it every day instead of all at once.

Huy (2001b) seems to agree. He suggests that the appropriate timing of multiple change interventions with different pacing characteristics is important because it helps to create a more tolerable and effective change rhythm. Periods of great change need to be balanced with periods of slowed activity. 'Excessive speed in changing lowers the organization's competence and leads to its collapse ... Each organization has to find its own dynamic internal change rhythm that permits it to alternate between rapid and moderately paced changes without losing synchronization and control' (Huy, 2001b, p. 610).

According to Abrahamson (2000), managers should intersperse major change initiatives with carefully paced periods of smaller, organic change allowing people some time to adjust. This 'dynamic stability' facilitates change without fatal pain. Resistance to change is frequently created by initiative overload and organisational chaos resulting from continuous large-scale change. Giving people the chance to 'tinker' with the existing system gives them some time to recover while still moving forwards. There are four rules of dynamic stability according to Abrahamson (2000):

1 Reward shameless borrowing as imitation is not a sign of weakness; it is a sign of the smart and efficient use of time and money.
2 Tinker internally first as many ideas that come from 'overseas', for example, are later rejected as not appropriate for 'us'. We already have intelligent people with great ideas within our organizations; listen to them and use them first.
3 Appoint a chief memory officer because we must learn from our mistakes and not re-invent the wheel.
4 Finally, hire generalists who are often derided as jacks-of-all-trades and masters of none, but whose range of skills lets them combine disparate ideas, techniques, processes and cultures. This leads to successful innovation and change.

Huy (2001b) proposes four ideal change intervention types that have implications for pacing:

Type 1: the commanding intervention approach (i.e., a commander-like approach whereby change agents apply directive and coercive actions to their change targets to exact compliance with their proposed change goals) is likely to be relatively effective at changing formal structures and should be used when change agents' purpose is to produce fast improvement in the firm's economic performance or when they value a quantitative conception of time and entrainment by factors outside the organisation and a time perspective that favours the near term.

Type 2: the engineering intervention approach (i.e., change agents' actions of analysing, understanding, and then redesigning work processes to improve the speed and quality of production) is likely to be relatively effective at changing work processes and should be used when change agents' purpose is to produce moderately fast improvement in the firm's economic performance or when agents value a quantitative conception of time and a time perspective that favours the medium term.

Type 3: the teaching intervention approach (i.e., an analytical and guided learning approach in which change targets participate in their own re-education through the active involvement of change agents; they are not passive, as in the commanding type: rather, targets collaborate in effecting their own personal change through changes in their fundamental beliefs) is likely to be relatively effective at changing beliefs and should be used when change agents' purpose is to develop the firm's organisational capabilities or when agents value qualitative inner time as a conception of time and a time perspective that favours a moderately long term.

Type 4: the socialising intervention approach (i.e., change agents' actions to enhance the quality of the social relationships among organisation members to realise organisational tasks; it is assumed that change in behavioural interactions among individuals will lead to change in beliefs and organisational culture) is likely to be relatively effective at changing social relationships and should be used when change agents' purpose is to develop the firm's organisational capabilities or when agents value qualitative social time as a conception of time and a time perspective that favors the long term.

According to Huy (2001b), starting large-scale change with the commanding type is likely to be effective in organisations that traditionally accept hierarchical authority, when the company has slack, and when change agents' power is concentrated. Commanding is likely to result in little resistance if it is done with benevolence, has a clear business logic that is acceptable to employees, and is done in a short time. Like Beer and Nohria (2000), Huy suggests that the commanding type of change must be followed by other intervention approaches to repair the social fabric of the organisation and improve work processes. It is also suggested that in organisations with little slack, low receptivity to radical change, dispersed power structures, or low innovation, starting large-scale change with the socialising, engineering or teaching types of change and ending with commanding is likely to constitute a more effective change sequence than starting with commanding. Both Beer and Nohria (2000) and Huy (2001b) suggest that combining seemingly opposite intervention approaches is likely to be accepted by recipients when change leaders apply a fair process and justify to recipients how these approaches are appropriate, and when leaders can complement, and coordinate well with, one another.

About 70 per cent of all change initiatives fail as managers try to do too much and lose focus (Beer and Nohria, 2000). This takes a heavy human and economic toll within organisations. Leaders need to recognise that their implicit models of change have significant implications for what will be achieved, how quickly change will take place, and what the costs will be. Beer and Nohria (2000) distinguish two 'ideal' models of change. Many managers attempt Theory E (similar to Huy's Type 1) changes based on economic value and usually involving heavy use of economic incentives, drastic lay-offs, downsizing and restructuring. E change strategies are more common where financial markets push corporate boards for rapid turnarounds. Theory O change (similar to Huy's Type 3) is based on the development of organisational capability, corporate culture and human capability through individual and organisational learning. Companies that adopt O strategies typically have strong, long-held, commitment-based psychological contracts with their employees. The question to ask is not which model of change is correct, but how can we get the best of both? The obvious way to combine E and O changes is to sequence them. Beer and Nohria's research suggests E change is best followed by O change. Even better, although very difficult according to the authors, is to adopt both types of change simultaneously. To do this, managers must explicitly confront the tension between E and O goals, not hide from them. Managers must set direction from the top and engage people below, plan for spontaneity by encouraging experimentation and evolution, as well as let incentives reinforce change, not drive it.

The last thing to say about the pace and types of change is that even if you have no control over them at all, there are still two things you can do to help reduce stress and re-energise people for change: (1) in addition to telling people what is going to change (i.e., who, what, where, how, why and when), be sure to tell people what is not going to change; and (2) celebrate success and progress to 'refill the bank'. Telling people about what is not going to change, as well as about what is, gives them something upon which they can count. It provides some sense of stability and continuity by giving people something they can hold on to, an anchor in the face of the storm of change. Celebrating and rewarding success and progress is something many change managers forget to do. Even though we sometimes have little control over changes that have been imposed on us from above or outside, we can exert some semblance of control by celebrating and rewarding achievement and effort. Recognition and rewards are powerful motivators and will be discussed in the next section of this chapter.

Existing Reward and Recognition Systems

In acknowledgement of their importance and power, much has been written on the topics of rewards and recognition. In organisations, to a great degree 'you get what you reward and measure'. The topic of measurement is discussed in more detail in Chapter 8. In this section, we will focus on rewards and recognition.

One of the problems with reward and recognition systems is that they are sometimes misaligned with our change-related goals. For example, if people are rewarded for individual achievement and their performance is appraised based on their individual achievement (e.g., in sales revenue, or the number of publications) while you are trying to implement some computer-supported cooperative work (CSCW) software or a team-based organisational structure, you are going to encounter problems as your current reward system is not aligned with the goals of your change intervention. CSCW requires cooperative behaviour in order to work and your reward system requires people to achieve individually if they are to be recognised and rewarded.

Lawler (2000) suggested that all too often a misaligned pay strategy not only fails to add value, but it produces high costs in the compensation area as well as inappropriate and misdirected behaviour. He suggested three possible solutions, as outlined below:

1 Pay the person. Person-based pay should be used to reward individuals for their skills, knowledge and competencies relative to their external market value. Compensation systems should be designed to suit individuals, rather than basing compensation on a position or job.

2 Reward excellence by paying for performance. Multiple pay-for-performance approaches should be used, with variable pay and stock as rewards. The pay-for-performance approach of an organisation needs to effectively translate its business strategy into measures that can be used for reward system purposes. One of the most difficult decisions in pay system design concerns whether to reward individuals, groups, business units or the organisation as a whole.

3 Individualise the pay system because one size 'does not' fit all when it comes to reward systems. Reward systems should be individualised to fit the characteristics of individuals that an organisation wishes to attract and retain. In most cases this can best be done by allowing individuals some choice regarding the rewards that they receive.

It is important to remember that not all reward and recognition systems must rely on money. Nelson (1994) offers these tips to follow during times of change:

1 Few things are as cherished by employees, especially during turbulent times, as knowing what is going on, and 'knowing' can take many forms. The more open and honest management is, the more employees will feel a part of the corporate community and whatever changes are imminent.

2 Share the organisation's purpose and mission and it will increase the likelihood that employees will identify with your goals. Explain current strategies and objectives for the year and help employees set goals that relate both to your goals and their own jobs.

3 When change is pending, take time to share what is changing to minimise the rumour mill and help alleviate fears and concerns. In the absence of accurate and timely information, people make things up. Explain how the change will affect the organisation, employees and their jobs. Ask for help to make the change easier. Do not always shove change down people's throats. This way, you will gain employee commitment to changes that occur.

4 Within 24 hours of each management meeting, share what was discussed and decided, and why. Take questions from employees and find out answers if you do not readily know them.

5 Ask employees what additional information they would like to have. The answers may surprise you.

6 Try to ask employees their opinions whenever possible, even if you are not always able to use the information they provide, and especially for topics that directly affect them. This basic courtesy goes a long way in helping employees to feel a part of things, and is a slap in the face when it does not occur.

There are two great articles that have been written by Steven Kerr in relation to this topic. In his classic article ('On the folly of rewarding A, while hoping for B', originally published in 1975 and more recently updated as Kerr, 1995), he outlined the following four general factors contributing to the prevalence of fouled-up reward systems:

1 Fascination with an 'objective' criterion. Many managers seek to establish simple, quantifiable standards against which to measure and reward performance. Such efforts may be successful in highly predictable areas within an organisation, but are likely to cause goal displacement when applied anywhere else.

2 Over-emphasis on highly visible behaviours. Difficulties often stem from the fact that some parts of the task are highly visible while other parts are not. For example, publications are easier to demonstrate than teaching, and scoring goals and hitting home runs are more readily observable than passing to team mates and advancing base runners. Similarly, the adverse consequences of pronouncing a sick person well are more visible than those sustained by labelling a well person sick. Team-building and creativity are other examples of behaviours which may not be rewarded simply because they are hard to observe.

3 Hypocrisy. In some instances, the rewarder may have been getting the desired behaviour, even though he or she claims the behaviour was not desired. For example, in many jurisdictions within the USA, judges' campaigns are funded largely by defence attorneys, while prosecutors are legally barred from making contributions. This does not do much to help judges to be 'tough on crime' though, ironically, that is what their campaigns inevitably promise.

4 Emphasis on morality or equity rather than efficiency. Sometimes consideration of other factors prevents the establishment of a system that

rewards behaviour desired by the rewarder. The felt obligation of many to vote for one candidate or another, for example, may impair their ability to withhold support from politicians who refuse to discuss the issues. Similarly, the concern for spreading the risks and costs of wartime military service may outweigh the advantage to be obtained by committing personnel to combat until the war is over.

In the second of Kerr's articles on this topic (Kerr, 2000), he identified ten principles related to best practice in rewarding performance:

1 Rewards should be the third thing an organisation works on, measurement should be the second, but clear articulation of desired outcomes should be the first.
2 If a reward is unavailable, do not try to use it.
3 If you make people ineligible for a reward, you take away their motivation to strive for it.
4 For rewards to be powerful, they must be visible.
5 If you want someone to perform, you should reward them when they do perform and not when they do not.
6 A long-deferred reward loses most of its power.
7 The best rewards are those you can take back if necessary.
8 Do not under-estimate the importance of non-financial rewards.
9 Get peers, subordinates and customers involved in your reward and measurement systems.
10 All principles have exceptions (except this one).

These principles are important and powerful. They are based on the psychological perspective referred to as behaviourism. This perspective views behaviour as the result of its perceived consequences. Behaviours that are rewarded therefore get repeated more frequently than those that do not. Rewards and recognition are important social-psychological structures that influence behaviour. In the next section, we more directly examine influence and persuasion.

The Science of Persuasion

The implementation process can be thought of as a persuasion and influence process. This entire book can be thought of as a complex persuasion process designed to help people acknowledge and overcome resistance so they can commit to, or at least comply with, the changes you are trying to

make. Leaders, change managers and anyone who wants to influence people in order to get things done must master the fundamental principles of
persuasion (Cialdini, 1985, 2001). The science of persuasion helps us learn
how to overcome resistance and increase commitment. Several questions
may arise at this point, including the following:

1 Why isn't logic enough to convince people?
2 Why can't the facts speak for themselves?
3 Why do we have to be concerned with trying to persuade and influence?
4 Can't everyone see how great this change will be and just do it?

The facts rarely, if ever, speak for themselves for several reasons. First,
what may seem like logic and/or fact to one person may not seem so to
another. Remember, behaviour is a function of both the person/personality
and the environment/situation: $B = f(P \times E)$. You have a certain personality
and background of experiences that enable you to think and 'see' the world
the way you do. A person with a different background, personality and value-
set who finds themselves in a different situation being influenced by a different set of expectations, roles, rewards and measures, will probably think
and 'see' the world differently from you. A second reason why the facts do
not usually 'speak for themselves' is that people tend to over-estimate the
importance and effectiveness of logical communication (Cialdini, 1985).
Logical communication does sometimes persuade, but it works best when
we like the person and feel involved in what is being said. The principles we
are about to discuss are not magic (they do not always work perfectly, either).
The basic premise here is that since logical communication is not enough to
persuade, what else can we try to increase our chances of success?

These principles are founded on basic human psychology. They are usually unconscious, fixed action patterns or mental short-cuts that affect our
judgements and perceptions and, therefore, our decisions. Blindly mechan-
ical patterns of action exist in a wide variety of species, including humans.
Consider the following example (adapted from Cialdini, 1985):

There are people waiting in line to use a library copier. In the first instance,
an individual walks up to the front of the line and speaks to the people stand-
ing there, 'Excuse me, I have five pages. May I use the Xerox machine?' In
this situation, people usually let the newcomer jump the queue about 60%
of the time. In other words, there is about 60% compliance to the request.
In the second instance, the situation is almost identical, but the individual
walks up to the front of the copier line and says, 'Excuse me, I have five

pages. May I use the Xerox machine because I am in a rush?' In this case, we get about 94% compliance to the request. In this case, by giving a reason or justification, 'because I am in a rush', we get greater compliance. More interesting, however, is the third case in which the individual walks up to the front of the line and says, 'Excuse me, I have five pages. May I use the Xerox machine because I have to make some copies?' Notice that the reason or justification given, 'because I have to make some copies' really adds no relevant information. Of course you want to use the copier because you have copies to make. Even though this is the case, this request generates compliance levels of about 93%, almost as high as in the second situation, and considerably higher than the first situation.

It is hypothesised that the act of justification, or providing a reason – 'I need to do X because' – acts as an unconscious trigger and elicits higher levels of compliance, regardless of the actual content of the communication.

Let us look at some of the fundamental principles of persuasion and influence that have been identified and researched, and how they can be applied to the strategic implementation of technology.

The Principle of Reciprocity

Reciprocity is a feeling of being obligated to future repayment. Many people naturally feel as if they must repay in kind. If I give you something, and you accept it, even if you did not ask for it or really want it, you will feel obligated to give me something in return. We see this principle being used in many ways. When free samples or memberships are given away at supermarkets and health clubs, for example, they are trying to capitalise on feelings of reciprocity. This principle also may be active in the relationships between pharmaceutical companies and medical researchers. According to an article in *The Australian* (Chapman, 2001), a *New England Journal of Medicine* study found that 37 per cent of medical researchers who published conclusions critical of the safety of calcium channel blockers had received prior drug company support, compared to 96 per cent of those who reported results supporting the drug's safety. Other studies have shown that in situations where straightforward mail requests for charitable contributions work 18 per cent of the time, when the mail requests are sent with a set of free, personalised address labels, even though the recipients did not ask for them, the requests work 35 per cent of the time (Cialdini, 1985). Including the free gift almost doubles compliance with the request. If you want to persuade, therefore, give what you want to receive. While gift-giving is the most obvious example of this principle, Cialdini reminds us that if managers want trust, they should first give it. If they want team-work and cooperation, they must first display it. 'Leaders should model the behavior they want to see from others.'

The principle of reciprocity also works if you have a big request that is rejected then followed with a smaller request. Studies have shown that if you stop random passers-by and ask for volunteers to help chaperone juvenile detention centre inmates on a day trip to the zoo, you get about 17 per cent compliance (Cialdini, 1985). If you stop another random sample and first ask if they would serve as a volunteer counsellor for two hours per week for the next two years, you get no compliance. If you then say, 'If you can't do that can you help chaperone juvenile detention centre inmates on a day trip to the zoo?', you get 50 per cent compliance. Many people already use this aspect of reciprocity when they ask for 50 per cent more money than they need for their change budget.

The Principle of Social Proof

We view a behaviour as 'more correct' in a given situation to the degree that we see others performing it. In a classic experiment, it was shown that if one person is looking up into the sky in a city, about 4 per cent of passers-by also look up. If five people are looking up into the sky in a city, about 18 per cent of passers-by also look up. If 15 people are looking up into the sky in a city, about 40 per cent of passers-by also look up. Another aspect of this principle is that people follow the lead of similar others. The lesson for organisational change managers is that persuasion can be extremely effective when it comes from peers. This also explains the power and value of pilot testing.

The Principle of Liking

People like those who like them. We prefer to say 'yes' to the requests of people we like. Studies have shown that we like those that are physically attractive (42 per cent versus 23 per cent success for attractive versus unattractive people soliciting charitable donations), similar to us ('No kidding. You majored in psychology? I majored in psychology too!'), pay us compliments (whether they be true or untrue), more familiar (those with whom we have more contact), and associated with something positive (conditioning).

If you want to persuade, uncover real similarities between you and the other, and offer genuine praise. This will increase the probability that they will like you, and your ability to influence them will increase.

The Principle of Consistency

Once people make a choice or take a stand (a verbal and/or behavioural commitment), we encounter personal and interpersonal pressure to behave consistently with that commitment. People align their behaviour with their clear commitments. Research has shown that when taking booking for a restaurant, if someone says, 'Please call if you have to change your plans', about 30 per cent of those who book do not show up. If, on the other hand, after taking a booking for a restaurant, someone says, '*Will you* please call if you have to change your plans?' and waits for an answer, the no-show rate drops to 10 per cent. Most people, once they take a stand or go on record in favour of a position, prefer to stick to it. The implications for the workplace and organisational change are to either 'get it in writing' or encourage a public declaration of intention if you want someone to follow a particular course of action.

The Principle of Authority

Perceived power and authority have a great effect. If a man in a suit and tie 'jaywalks' (i.e., crosses the street when the traffic signal is red), compared to a casually dressed man, about 350 per cent more people follow. People defer to experts. When trying to implement innovation and technical change, therefore, if you want to persuade, expose your expertise. Do not assume it is self-evident.

The Principle of Scarcity

People want more of what they can have less of. People assign more value to opportunities when they are less available: 'If you do not do it now, you may never have the chance again.' People frequently assume that the rare is beautiful. In order to influence more effectively, therefore, you must highlight the unique benefits of your product, service or idea, and the exclusivity of the information you possess.

The Principle of Emotion

Our cognitive processes are frequently suppressed by our emotional reactions (e.g., fear, jealousy, greed, joy, love). As change managers, an emotional appeal to act in a certain way (e.g., an appeal to courage and self-sacrifice; an appeal to save our families, children, jobs, the company, the country, etc.) can have a powerful effect. This explains why the 'burning platform' frequently motivates change.

The Principle of Imitation

By imitation, we mean modelling a new behaviour or being a role model so others can imitate your behaviour. The days of being able to say, 'Do as I say, not as I do' in organisations are long gone. If you want people to be flexible and to change, you must be willing to do the same. If you lead the way with your behaviours, others are more likely to follow.

The Principle of Perceptual Contrast

This principle is one of my favourites. What you are thinking of just before and while you are making a decision or judgement affects that decision or judgement. For example, if you are thinking of a relatively small number (e.g., the price of a newspaper or the price of some lollies or gum) while you are trying to estimate the cost or length of something truly unknown to you, you are likely to make a lower estimate than if you are thinking of a relatively large number (e.g., the price of a new home) just before you make your judgement. The numbers you were thinking about just before you make your judgement 'anchor' you lower or higher.

CHAPTER 7

Roll-Out

With the ground as well prepared as possible, it is time to plant our seeds. Now that people, groups and social systems have been considered, we should move our attention on to more traditional 'technical' project management activities and to roll out the new technology. By this point in the implementation process as much of the preparatory work as possible is complete and as many people-related issues as possible have been dealt with. Plans are now executed and the transition process from the 'present state' to the 'desired state' is managed. This is the stage that most people equate with implementation, and it is when many of the planned changes actually take place. Many of the group/individual-level implementation activities from Chapters 4, 5 and 6 carry on into this stage and the activities associated with this technical roll-out stage seamlessly flow into the confirmation stage (Chapter 8). In this chapter, we will briefly consider the importance of technical training, some options for systems conversion and project change management.

Technical Training

Once you have addressed the issues in the previous chapters you should be ready to actually begin technical installation of the innovation or technical changes. You will have completed initial education and training of the appropriate people, made them aware of what is going to happen, put in place any necessary structures, policies and procedures, and helped people reach the point of committing to, or at least being convinced of, the fact that implementation will occur.

As mentioned in Chapter 4, education and training are so important they need to happen at three different times during the implementation process. During the early parts of the implementation programme people will need general awareness-raising regarding what is going to happen, as well as education and training in broad areas such as communication skills and working in groups. This roll-out stage is when the second wave of training must begin. This roll-out-related training must be more operationally focused and specific to the new technology, new strategy or innovation being implemented. Some time after this, the third round of education and training needs to be conducted. It should be focused on issues related to how to use the more sophisticated or advanced elements, features and functions of the system and/or software and/or how best to take advantage of the innovation, now that there has been time to deal with the basics.

System Conversion

Converting systems is a critical 'technical' task that must take place at about this point in the process. According to a popular management information textbook (Laudon and Laudon, 1996), conversion is the process of changing from the old system to the new system and answers the question, 'Will the new system work under real conditions?' Whether your system is a piece of software, a management information system, or a new structure, strategy or work organisation, you still have to change from the old system to the new system.

This is the point when the old system is turned off and the new one turned on. It is the point when the transition is made from the old work arrangements to the new team-based structures, for example. At this point you can finally attend to the logistical processes of installation. You can execute your plans to deliver and set up the innovation or new technology. You may be converting prototypes into fully operational systems or switching from old systems to new ones. If it is a social or administrative innovation, it is time to move from the small-scale, pilot-testing stage to a full roll-out of the processes.

While there are any number of different ways to make these transitions, there are three broad conversion options that should be considered. Each option has its pros and cons and some options are not possible with certain innovations or new technologies. The first option is the slow migration from old to new. This conversion option gives you time to learn from early trials and to make the necessary adjustments. Unfortunately, it takes longer than a quick switch-over, and may be seen as evidence of uncertainty. This option, however, has the benefit of allowing users time to adapt to the changes. It is important sometimes to pace the introduction of change to allow people the chance to 'catch up' and absorb what has happened before looking forward to more change. We addressed this issue of pacing in Chapter 6.

The second conversion option is to run parallel systems. In a parallel conversion strategy, both the old system and its potential replacement are run together for a time until everyone is assured that the new one functions correctly. Running parallel systems allows you to have a back-up as you are trying out the innovation or new technology. This is a particularly good option if the continual operation of the systems is critical. Unfortunately, it frequently costs a great deal of money to run redundant systems and sometimes it may not be feasible technically.

A third conversion option is to make a quick switch from the old to the new system. Although this option is decisive and may be more cost-effective in the short run than operating parallel systems or a slow migration, there is a danger in terms of reduced operational effectiveness. If the conversion is not successful it could cause major problems. If reliability is not an issue, or if your systems are not rigid, then you could consider this conversion strategy.

Regardless of which option you choose, there are several other conversion issues to consider, as shown below:

1 Budget sufficient time, money and effort for conversion, especially data conversion.
2 Educate and involve everyone who will use the innovation or new technology before installation.
3 Begin training before the innovation or new technology is actually delivered and ready to be 'turned on'.
4 Do not make the innovation or new technology operational before it is fully ready. Even if you are over time and over budget, rolling out a 'product' while it is still incomplete will only make it worse.
5 If appropriate, develop quality documentation that is written from the users', not the developer's, perspective.
6 Provide people, money and time for proper evaluation, modification and maintenance.

Project Change Management

Kezsbom and Edward (2001) suggest that modern project management is a process that offers the contemporary organisation a unique vehicle for change. It is a methodology through which we can create more flexible, adaptive, yet accountable corporations, without being bound by the more traditional policies and procedures originally designed to maintain the status quo. Global competition requires managers, leaders, and professionals to think of ways to change their organisation continuously to gain competitive advantage. No longer can projects, large or small, be run by 'seat of the pants' management practices, and neither can traditional project management practices, originally conceived to generate and control repetitive tasks within hierarchical structures, meet the demands of today's politically sensitive, rapidly changing environment. Project management processes must be tailored to accommodate the unique, dynamic and diverse needs of varied projects, while maintaining control over cost and quality. Modern project management processes can meet these unique challenges. As Frame (1994) states:

> Traditional project management has enabled humans to do some incredible things. For example, it provided the U.S. National Aeronautics and Space Administration (NASA) with the management capability to put men on the moon. It makes possible the construction of oil drilling platforms in the North Sea. It provides airplane manufacturers with the discipline to design and build complex commercial aircraft ... The problem is that traditional project management is broken.

Traditional project management is overly focused on technology and is populated by technical 'experts' who often do not understand the end-user's or customers' perspective, and 'who design and build products that are of personal interest to them. They are often driven to build things that will gain them the admiration of their fellow experts' (Frame, 1994, p. 5). In other words, traditional project management is broken because it often overlooks the social and people/customer/user perspective. In order for project management to be effectively employed in the business environment of today, it must first become more customer/people-focused, it must explore the use of new management tools, and it must redefine the role of project managers, giving them more power to operate effectively (Frame, 1994).

Traditional project management emphasises acquiring basic skills in scheduling, budgeting, and allocating human and material resources (Frame, 1994). These are the primary tools of project managers who are mere technical installers. In their new expanded role, project managers and project staff need a different set of skills to be effective. Project managers and staff should be proficient in such 'hard' skills as the basics of contracting, business finance, integrated cost/schedule control, measuring work performance, monitoring quality, and conducting risk analyses; they should also be adept at such 'soft' skills as negotiating, managing change, being politically astute, and understanding the needs and wants of the people they deal with (e.g., customers, peers, staff and their own managers).

Exhibit 7.1 Technical project management and roll-out

The 10 elements of project management (adapted from the Project Management Institute's [http://www.sviamerica.com/syscon.html] PMBOK Guide, 2000; Quality management – Guidelines to quality in project management ISO 10006:1997(E); and Harrington, Conner and Horney, 2000) are given below.

1 Project integration/interdependency management (project initiation, project plan development, interaction management, overall change control/change management, closure).

2 Project scope management (initiation/concept development, scope of planning, scope definition, scope verification, scope change control, activity definition and control).

3 Project time management (activity definition, activity sequencing and dependency planning, activity duration estimating, schedule development, schedule control).

4 Project cost management and (resource planning, cost estimating, cost budgeting, cost control).

5 Project quality management (quality planning and, quality assurance, quality control).

6 Project human resource management (organisational planning, organisation structure definition, staff acquisition and allocation, team development).

7 Project communication management (communication planning and control, information management and distribution, performance reporting, administrative closure).

8 Project risk management (risk identification and estimation, risk quantification, risk response development, risk control).

9 Project procurement management (procurement planning/resource planning, solicitation planning, solicitation, contract administration, contract close-out).

10 Managing organisational change to … (change planning, define roles and develop competencies, establish burning platform, transformation management).

You can use the sites listed below to find out information on the latest in project management techniques and tools. Search the web for project management software and techniques. Did you know that you could rent project management software on-line? You may like to try to find some sites that provide this service.

URL	http://www.fek.umu.se/irnop/projweb.html
Title	Umeå University, Department of Business Administration
Comment	Dr *Johann Packendorff*, Umeå School of Business and Economics, Sweden maintains a guide to project management research sites including professional associations, networks, journals, researchers and more.

URL	http://www.mtgi.com/
Title	Management Technologies Group website.
Comment	'Explore [their] web page to get practical information, ideas, tools, and techniques on project management planning, scheduling, cost estimating, training, team building, professional development and project management trends and developments.'

URL	http://www.linkbank.net/get_links/default/CBPKnowledgeBank
Title	The Center for Business Practices, KnowledgeBank Links
Comment	Search best practices, government, information technology (IT), research conferences, publications and more associated with management in general and project management in particular.

The Roles and Skills of Project Managers

If an IT recruiter could write an ad for the perfect project manager, here is what it might say (Rice, 2001):

> Wanted: Highly organised techie who likes people and calendars as much as his/her computer. Must be obsessed with sticking to a schedule and following a budget, but not so rigid that s/he drives his/her team members crazy. No antisocial coders looking for a pay raise need apply. Generous salary and responsibility.

Given such requirements, it is no wonder that good project managers are some of the toughest IT workers to find. Few people have such a balanced combination of technical, interpersonal and business skills. Harrington, Conner and Horney (2000, pp. 146–7) agree that good project managers need to be multi-skilled and able to handle both the technical and people/social side:

> Project managers must be as skilled and familiar with managing organi-sational change concepts as they are with financial management. In fact, the skills that project managers have related to managing organisational change often have a much greater impact upon the outcome of the proj-ect than financial controls. The degree of resistance to change impacts on cost, schedule, resource requirements, and the performance of the end output from the project. No longer can project managers limit their proj-ect design to just the resources consumed by the project. An effective project management plan must prepare the targets (the people who need to change) so that the results of the project will be effectively assimilated into the organization.
>
> Too often, project managers look at the four key project management factors – process, knowledge, technology, and people – and limit their people thinking to the composition of the project team. As a result, the project team is made up of technology experts who have little or no knowledge or concern about the people who are affected by the project and who have to live with the project's results day after day.
>
> The project manager must be skilled at:
>
> - being a change agent
> - being a change advocate
> - being a change facilitator
> - being a change target.

In other words, good project managers must be concerned with both the hard (e.g., technical and financial) as well as the soft (e.g., motivation, communication, relationships) sides. In fact, Rice (2001) suggests that if you are the technical type that would rather 'do it yourself' to get something done or solve a problem rather than manage others to do it, project management is not for you. Rice (2001) quotes several managers who echo these sentiments:

'I once had a project manager, who'd been a very good engineer, tell me that, if it weren't for all the interruptions and talking to people, he could just go back to his office and solve the problem himself.' ... If you can relate to that engineer's frustration, hard-core IT probably is where you should stay – but if you love working as a team with your coworkers, developing plans and sticking to a schedule, perhaps project management is a career track to pursue.

[A project manager's] responsibilities tend to be more administrative and people-oriented. While a technical lead provides the necessary 'techspertise', the project manager spends his day checking in with anywhere from a handful to hundreds of team members about the project's progress. He communicates the client's needs to the programmers and software developers. He enforces their deadlines. He may monitor costs, to ensure the work is done within budget. But a project manager works with more than the techies. He also serves as a liaison to the people who ordered the project, whether that's an internal 'customer' such as the accounts payable department or an external client such as an advertising firm. It makes for a heck of a lot of phone calls and e-mails. Eighty percent of the job is communication.

But that emphasis on communication and other soft skills doesn't mean that project management is the perfect high tech job for those with few, if any, tech credentials. Many project managers have at least a few years of IT experience ... You need to be familiar with the technology you're using, the folks reporting to you will rely on getting good feedback from you.

Variety is one of the best parts about [the] job. I can touch every part of a project I like: the numbers, the mixing of people, and the interaction with the customer. It's a different challenge every day ... Sometimes, those challenges mix the emotional and professional, as team members clash or a client gets upset. By the nature of project management, there's a lot of conflict. If a person doesn't like conflict, it's the wrong field to get into.

Frame (1994) has conducted an informal survey of hundreds of project managers over a number of years trying to identify the traits of an ideal project manager. The following list enumerates general traits that have emerged from this survey. The ideal project manager should:

- have a thorough understanding of the project goals
- be capable of understanding staff needs
- have a good head for details
- have a strong commitment to the project – that is, be willing to put in long hours on the project
- be able to cope with setbacks and disappointments – in the discipline whose guiding principles are governed by Murphy many hours are going to be anxiety ridden
- possess good negotiation skills since a large part of project life will be spent trying to acquire resources
- be results-oriented and practical
- be cost-conscious and possess basic business skills
- be politically savvy (i.e., be aware of what not to do as well as what to do)
- have a high tolerance for ambiguity since little is clear on most projects

According to Kezsbom and Edward (2001), traditionally project managers have been considered as technical experts. They made most, if not all of the decisions, and defined the job and how it was to be done. Teams were formed only when needed, tasks were specialised and organisation was hierarchical. Now project managers are better educated, more open, friendly and people-oriented. They are better listeners, more quality-conscious, more participative and receptive to new ideas and the ideas of others. They are more involved in socio-emotional communication, they allow more independence and encourage greater cross-functional interaction. In the future, project managers will need to be facilitators skilled at group process and group dynamics and able to encourage others to participate in plans and decisions. They will need to understand how to coach, inspire and motivate. They will no longer be perceived solely as the technical expert. They will be boundary spanners who are comfortable relying on the expertise of others.

Many large IT or construction projects are run by professional project managers who have no other assignments. With smaller projects, however, the individuals who serve as project managers often spend only a small percentage of their time managing a project or they manage several at once. In either case, these project managers are part-time and their project management duties are frequently added on top of their normal roles.

This argues for the professionalisation of the role of project manager and for recognition that the role takes a great deal of time, energy and effort. If you want your change projects to be a success, train project managers and give them the time and resources they need to get their job done.

How Organisational Change Fits into Project Management

One of the problems with the strategic implementation of technology is that we often do not take the project management of a change as seriously as we take the project management of the building of a bridge or the development of a complex piece of software. According to Harrington, Connor and Horney (2000, p. 95), without an effective integration of project management techniques and change management implementation planning techniques, we are left with multiple plans (e.g., an organisational design plan, a change management plan, a communication plan, and a training plan). Attempting to manage multiple plans without clear milestones and integration points is difficult at best and does not lead to effective projects. Such unintegrated plans result in change management being bolted on as an afterthought. The 'bolt-on' approach to dealing with the human aspects of change management generally occurs only when the project runs into trouble. As the project manager confronts poor sponsorship, high levels of resistance, cultural conflict, and/or a lack of change management expertise through the project, there is a last-minute rush to attach a few change management solutions in a vain attempt to get things back on track. This contributes to project failure and results in no change or change that is superficial, short term or distorted.

Consider the model of the integration of the project management and change management processes shown in Figure 7.1. The model depicts the organisation as an open system taking inputs, transforming them and producing outputs. There is, of course, feedback from the outputs back to both the inputs and the transformation systems. Within the transformation system we see the interaction of the project management and change management processes, with the 'normal' production and service delivery organisational transformation processes. The model is meant to illustrate that an organisation is a complex system and that as it adds value (i.e., transform inputs into outputs) it can benefit by considering project management and change management as important value-adding components.

Harrington, Conner and Horney (2000, adapted from pp. 162–4) suggested the following tasks would be undertaken when change management

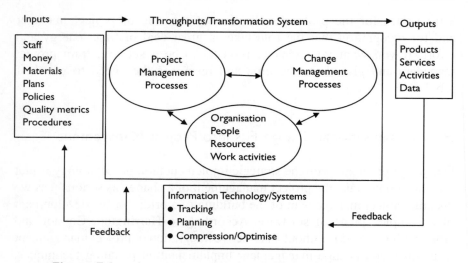

Figure 7.1   A systems approach to project change management

Source: Adapted from Kezsbom and Edward (2001), p. 12

is integrated with a complex project:

Project start-up and preparation

- Initiate project
- Define project charter
- Clarify scope of organisational change
- Develop project plan
- Review and approve project charter and project plan
- Kick off the project

Current people infrastructure description (similar to our knowledge and awareness-raising activities)

- Describe the current organisational environment
- Define current change management process
- Review and approve current people-infrastructure stage

Enterprise people infrastructure definition (similar to our facilitating structures)

- Define people enablement framework
- Review and approve enterprise people infrastructure stage

Pilot planning

- Assess change management enablers and barriers (resilience, change knowledge, managing adaptation resources, sponsor commitment, target resistance, cultural alignment, change-agent skills)
- Define pilot phase strategies
- Develop pilot phase transition management plan
- Develop initial pilot phase schedule
- Develop initial project charter
- Review and approve pilot planning stage

Transition management

- Refine transition enablers and barriers assessment
- Refine transition management infrastructure
- Develop transition management plan
- Review and approve transition management stage

Future state design

- Perform people-enabler detailed analysis
- Define people-enabler design
- Review and approve future state design stage

People enabler development

- Develop people enablers

Training development

- Develop training programme
- Prepare for training
- Review and approve training development stage

Business case refinement

- Conduct cost/benefit analysis
- Analyse risk and return
- Review and approve pilot business case refinement stage

Implementation planning

- Develop implementation plan
- Develop initial project charter

- Review and approve implementation planning stage

Communications management

- Implement communication plan
- Develop communications status report
- Conduct a change communication intervention
- Refine implementation management infrastructure
- Review and approve communications management stage

Staff training

- Conduct training
- Review and approve staff training stage

Change implementation monitoring

- Track and monitor enablers and barriers management plan
- Review and approve change implementation monitoring stage

Evolution planning

- Verify business value
- Identify evolution options
- Develop evolution vision
- Develop evolution plan
- Review and approve evolution planning stage

Project review and assessment

- Review project performance
- Close down project

Some Common Project Problems

Before we conclude this chapter by looking at some common project problems and a solution to some of them, let us look at an excerpt from Cleland (1998). Cleland is one of the foremost leaders in project management theory and has been referred to as the 'father' of project management (Cabanis-Brewin, 1999; Project Management Best Practices Report, www.pmsolutions.com/cbp, October, vol. 1, no. 1). According to Cleland (1998, ch. 19, p. 441):

> To be competitive today, a company has to practice continuous improvement and be nimble in responding to changes in markets and in the technology of products and processes. Two potential strategies exist for companies to pursue in advancing the technologies of their products and processes: (1) continuous improvement of the products and services that are offered and (2) a push for a 'breakthrough' in technology of products and processes. Both strategies are facilitated by the use of project management techniques. Which strategy to pursue depends on the resources available to the organisation and the amount of risk that the managers are willing to assume. To do nothing means obsolescence and ultimate dissolution.

In this single paragraph, he ties together several of the topics from this chapter and foreshadows some of the issues we will consider in the next.

While all projects are different, Schaffer (1997) has identified five fatal flaws that can kill any project, as shown below:

1 Projects are defined in terms of consultants' contributions or products (and not in terms of specific client results to be achieved). Instead, projects are designed to achieve some tangible results not just to install a system, produce a report, deliver some recommendations, or create strategic options. If the goal is quality improvement then some measurable improvement in quality must be defined as the target of the project. If lower service delivery costs are the end goal of the change, then measurable cost reduction in service delivery must be the key goal of the project is well.

2 The project scope is based on subject-matter logic and not on client readiness for change. By ignoring client readiness, there is often a gap between project recommendations and what the client is ready and able to do. If the focus of the project becomes to install a system or produce a report, the project managers naturally become preoccupied with doing the installation or the analysis and the report and they forget that the organisation must actually be able to accept and carry out the recommendations of the report or use of the new technology.

3 A one-big-solution design is used rather than incremental successes. Because of this, projects tend to be long-term, expensive and require considerable upfront investment. Instead, it is best to carve off some rapid-cycle projects that seem reasonably certain of success.

4 A hands-off approach is taken instead of working in partnerships. Because of this, the project team will be doing most of the work and integration with the business unit will not be planned for. This means those ultimately responsible for using the system will not be engaged in the project and will learn little. Instead, the two parties should be collaborating in a partnership model.

5 There is labour-intensive use of consultants (instead of leveraged use). This results in projects that are too long, too complex and too expensive.

Walker (2001) suggests that applying a traditional risk-management structure to project management can help reduce many problems which plague change projects. If we work in organisational cultures that reward results, most of our focus will be on results, not on managing the process. Frequently, project planning is based on limited information and a lack of research because of the need to meet project deadlines. In the initial excitement of a new, potentially profitable venture we sometimes make poor decisions. For example, the perceptions of many project initiators are positively biased, and this affects their overall perception about the project. As a result, they may be less likely to consider the project risks. Walker (2001) discusses project risks in three areas.

- *Risk Identification*
Building excitement over the potential of a new project is desirable to motivate those whose participation is needed to make it happen. Do not discourage excitement! However, we should not allow the excitement of a new project and its promised benefits to deter serious and systematic consideration of the risks associated with that project. Neither should we allow the immediacy of a project to justify shortcutting the risk assessment process.

- *Risk Evaluation*
There are many risks and opportunities associated with projects that are relatively unimportant and that would not significantly affect their success. There are other risks and opportunities, however, that can 'make or break' a project. No one can be reasonably expected to plan perfectly or consider every eventuality. By prioritising risks and opportunities in some way, however, we can better manage those 'make or break' issues.

- *Risk Perception and Politics*
Despite what some practitioners may believe, risk assessment is not an exact science. Science can be applied as far as a methodology to ensure consistency of application, consideration and evaluation of data, presentation of

data and so on, but there are always intangibles in the mix that can significantly affect the results of a risk assessment exercise. At least two of those intangibles are risk perception and politics. Risk perception is a complicated issue; suffice to say for our purposes that everyone involved with a project has his or her own perception of what risks exist and the extent of their importance. An individual's perception of risk has largely to do with that person's life experience and field of expertise. Their perception is reality to them. Therefore, each person's input regarding risk is important and should be considered. How do we do that in a way that is not too cumbersome and yields a satisfactory conclusion? It is a dilemma that cannot be fully solved but can be minimised by the application of proven, systematic risk assessment tools and techniques and by observing guidelines for group risk assessment processes:

> Whether it is a project to improve safety in your operations, to develop and market a new product, to build name recognition of your company, or to acquire another company, a systematic, comprehensive, risk-based approach is the only way to determine the risks of that project and to decide whether they are prudent. (Walker, 2001, p. 163)

CHAPTER 8

Confirmation: The Measurement of Change

Once the technology has been rolled out, it frequently goes through a process of testing and evaluation during which decision-makers and users attempt to confirm, modify or reject the choice. Various measurement issues must be considered taking a balanced approach including both hard (e.g., financial, uptime) and soft (individual satisfaction, group morale and functioning) measures across individual (customer and employee), group, organisational and societal levels of analysis. Eventually, the successful new technology or innovation gets incorporated, both technically and socially, into 'the way we do things around here'. It becomes the norm and part of the routine.

In this chapter, therefore, we will first discuss the importance of measurement for the success of strategy and technology implementation. We will then look at the issue of what to measure and how we need a number of balanced measures that link strategic and operational objectives. Finally, we will address the issue of when to measure and the importance of feedback.

The Importance of Measurement

The first thing to say about the measurement of organisational change is that we have known for years that it cannot be done perfectly. Unlike experimental evaluation done under controlled conditions, the evaluation of real-world organisational change suffers from multiple purposes, overlapping and interdependent interventions, and many other difficulties associated with 'field', as opposed to 'laboratory', conditions.

Since we cannot do it 'correctly', why should we do it at all? There are several good reasons to evaluate organisational change, even though it must be done imperfectly. Increasingly we are being asked to justify the high costs (both direct and indirect) of change interventions with hard data. Since organisations are under great pressure to cut costs and do more with less, this trend is likely to continue. We are asked if the ends justify the means, if the outcome was achieved, and was it worth the effort and expense? We must therefore be able to account, even if it is an imperfect accounting, for the success and the costs of change. It is better to be vaguely right than precisely wrong.

Another reason to try to measure organisational change, even though we cannot do it perfectly, is because research indicates that organisations which measure are more likely to succeed. Research suggests 'measurement plays a crucial role in translating business strategy into results ... You simply can't manage anything you can't measure' (Lingle and Schiemann, 1996, p. 56). In fact, data show organisations that are tops in their industry, stellar financial performers and adept change leaders distinguish themselves by the following five characteristics (Lingle and Schiemann, 1996; Schiemann and Lingle, 1997):

1 They have agreed-upon measures that managers understand.
2 They balance financial and non-financial measurement.
3 They link strategic and operational measures.
4 They update their strategic 'scorecard' regularly.
5 They clearly communicate measures and progress to all employees.

These five characteristics will provide us with the structure for the remainder of this chapter and each will be discussed in the subsections that follow.

What Should We Measure?

The first characteristic of effective measurement-oriented organisations, according to Lingle and Schiemann (1996), is that they have agreed-upon measures that managers understand. Managers in these well-measurement-managed organisations know the answer to the question: 'What are we going to measure: the changes/progress/the process of change/the outcome(s) of the change?'

This highlights one of the most basic distinctions in the design of evaluation research. Formative or process evaluation designs concentrate on providing regular, systematic feedback to programme designers and implementers so they can modify the programme on an ongoing basis (Legge, 1984). Summative or outcome evaluation, on the other hand, is concerned with identifying and assessing the value of programme outcomes in the light of initially specified success criteria, after the implementation of the change is completed (Legge, 1984). In other words, are we going to measure our change process and progress and/or are we going to measure the outcomes?

Besides measuring the process and the outcomes, there is, of course, the potential to measure the inputs to the process. Some recent work, by two of my colleagues at the Australian Graduate School of Management, provides a framework that may help us clarify our thinking on this issue. Consider the excerpt from Vlasic and Yetton (2002) given in Exhibit 8.1.

Exhibit 8.1   Delivering successful projects: research questions

Project Uncertainty

Uncertainty is one of the key determinants of performance (Locke and Latham, 1990). As uncertainty increases, the probability of success decreases. In identifying uncertainty, a number of authors have developed lists of what they see as the key dimensions (Boehm, 1991; Schmidt, Lyytinen, Keil and Cule 2001). These lists examine broadly the same areas, including individual, project, organisational and environmental characteristics.

How do firms manage uncertainty? Numerous studies have examined mechanisms to reduce uncertainty, such as top management support, user involvement, prototyping, project plans, progress monitoring, resourcing and cross-functional teams (Lyytinen, Mathiassen and Ropponen, 1998). Although there is no doubt that a combination of these is important, there is little consensus in the literature on how they relate to one another. For example, Deephouse, Mukhopadhyay, Goldenson and Kellner (1995–96) report that, of all the mechanisms proposed by researchers to reduce uncertainty, only project planning and cross-functional teams had significant influence. What is absent in this stream of research is consensus around a strong analytical framework within which to examine project performance. Organisational control theory is such a framework. It is used elsewhere to analyse and explain task performance (Eisenhardt, 1985). Here, we adopt and apply it to project management.

Organisational Control Theory and Project Uncertainty

Organisational control theory is concerned with everything that helps people in a firm attain organisational goals (Eisenhardt, 1985; Ouchi, 1979). It assumes tasks can be managed by focusing on three types of mechanisms: specifically, input, behaviour and output control. Input control is how the materials and human resource elements of a task are managed. Behaviour control includes process rules and procedural norms. Output controls are performance measures such as cost, time and customer satisfaction. Few studies have been conducted in the information technology (IT) literature using control theory. The exceptions are Henderson and Lee (1992), Kirsch (1996, 1997), Kirsch and Cummings (1996) and Liu, Yetton and Sauer (2001).

In assessing the use of control mechanisms, control theory identifies two independent contingent variables, outcome measurability and task programmability (Eisenhardt, 1985), which are defined as follows.

1 *Outcome measurability* is an outcome's susceptibility to reliable and valid measurement. Outcomes can be measured in terms of time, cost, quality and scope.
2 *Task programmability* is a task's receptiveness to a clear definition of the behaviours needed to perform them. Tasks are the activities to be performed on a project.

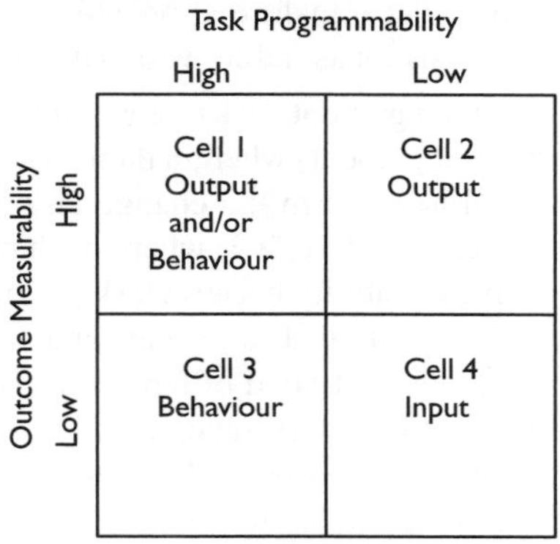

Task Programmability

Figure 8.1   Control mechanisms

These two variables help to determine the level of uncertainty in a task. Figure 8.1 presents the relationship between the choice of the appropriate control mechanisms and the task context, as defined by these two dimensions. If firms use the control mechanisms that match the context, they increase the probability of success. For example, in Cell 1, firms have a good idea of the tasks and the outcomes of those tasks and so are recommended to use output and/or behavioural controls. Conversely, in Cell 4, where tasks and outcomes are uncertain, input controls are more appropriate.

In addition to the performance benefits of matching the control mechanisms to the context, research has also found that firms that are in Cell 1 commonly perform better than firms that are in Cell 4 (Liu, Yetton and Sauer, 2001). This finding is consistent with goal-setting theory which, given that Cell 4 is a more uncertain context than Cell 1, predicts that performance would be higher in Cell 1 (Locke and Latham, 1990).

Examples of tasks and industries that fit into each of the Vlasic and Yetton cells, for the purpose of our conversation regarding the strategic implementation of technology, are:

Cell 1. Manufacturing tasks and construction tasks wherein you know what you are doing/delivering and how to go about it, and outputs of the process are measurable. The task is programmable (there is a clear definition of the behaviours needed to perform the task) and measurement of the outcome in terms of time, cost, quality and scope is valid and reliable. Since the transformation process is knowable and the outcomes measurable, you can use either behavioural control of the process or output control.

Cell 2. Events management tasks (e.g., Olympics) and tasks in many sales organisations (e.g., insurance sales, advertising agency, clothing boutique) fall within this cell. You can measure the outputs (e.g., sales volumes), but you do not know exactly how to go about it. The task is too complex and variable to be programmed (i.e., no clear definition of the behaviours needed to perform the task) but measurement of the outcome in terms of time, cost, quality and scope is valid and reliable. Since the transformation process is not well known and agreed upon but the outcomes measurable, you can only use output control.

Cell 3. Simple change management tasks (e.g., the change from traditional offices to an open office arrangement) wherein there is general agreement regarding the behaviours needed to perform the change task but measurement of the outcomes is not valid or reliable. In cell 3, where we have moderate uncertainty mainly around outcome measurability because task programmability is high (i.e., there is a clear definition of what needs to change and how to do it) but outcome measurability is low (i.e., low likelihood of any valid and reliable measurement of the outcome in terms of time, cost, quality and scope of the change), only behavioural control of the process is possible. Behavioural control, in our case, is control over the change process. In other words, to ensure the success of relatively simple change and the highest quality outcomes possible, we should focus on ensuring that the change process itself is the best it can be. Since there is good agreement regarding the behaviours needed to perform the change task, but the likelihood of valid and reliable measurement of the outcome of change is low, the things we can control are within the change process itself. The better the process, the better the outcome, even though we cannot get satisfactory measures of the outcomes. In this case, it seems we will be better off focusing on formative or process measures so we can provide regular, systematic feedback to implementers so they can modify the change programme on an ongoing basis.

Cell 4. Complex organisational change tasks (e.g., the implementation of strategy or complex IT systems implementation) wherein there is neither general agreement regarding the behaviours needed to perform the task nor valid and reliable measurement of the outcomes. When task programmability and outcome measurability are both low, we have maximal uncertainty. Since we cannot measure the outcomes, we agree on the process. The only thing we can control are inputs to the process, such as how the planning, material and human resource elements are managed. Translating that into the language we have been using regarding the implementation of strategy and innovation, it means that to ensure successful complex IT or strategic change and the highest quality outcomes possible, we should focus on ensuring that the inputs to the change process, such as education and training, selection and placement, project planning and so on are the best they can be. Since there is no general agreement regarding the behaviours needed to perform the change task, or the likelihood of any valid and reliable measurement of the outcome in terms of time, cost, quality and scope of the change, the only things we can control are the inputs to the process. The better the inputs, the better the outputs even though, once again, we cannot get satisfactory measures of the outcomes. In this case, it seems we will be better off focusing on measuring the quality of the inputs, as, by definition, summative or outcome evaluation is not possible.

While outcome evaluation is our preference, or our ideal (as this allows us to identify and assess the value of a change programme in the light of initially specified success criteria after the implementation of the change is completed), the work by Vlasic and Yetton (2002) suggests that when this is not achievable, we should try to measure outcomes as best we can, and we must try to control and measure inputs. Consider the following from Ouchi (1977, p. 98) regarding input controls:

In the final cell [cell 4], where the transformation process is not known and outputs are unmeasurable, only ritualized control is possible; it is similar to a situation in which no learning can take place because correct behaviours and outputs cannot be identified. Thus, whatever attempts at control take place are supported by ritual rather than by what we ordinarily think of as rational analysis. These rituals may have the effects of providing the appearance of rationality and therefore legitimate the activities of what is really a quite loosely coupled organization. While not wanting to dwell on this condition, organizations in this cell tend to rely heavily on the selection process as their only means of effective control.

Similarly, he later discusses input controls as being like 'clan' control via ritual and ceremony (Ouchi, 1979, p. 844):

[In cell 4] ... suppose that we are running the research laboratory at a multibillion-dollar corporation. We have no ability to define the rules of behaviour, which, if followed, will lead to the desired scientific break-throughs, which will, in turn, lead to marketable new products for the company. We can measure the ultimate success of the scientific discovery, but it may take ten, twenty, or even fifty years for an apparently arcane discovery to be fully appreciated. Certainly, we would be wary of using a strong form of output control to encourage certain scientists in our lab while discouraging others. Effectively, we are unable to use either behavior or output measurement, thus leaving us with no 'rational' form of control. What happens in such circumstances is that the organization relies heavily on ritualized, ceremonial forms of control. These include the recruitment of only a selected few individuals, each of whom has been through a schooling and professionalisation process which has taught him or her to internalize the desired values and to revere the appropriate ceremonies. The most important of those ceremonies, such as 'hazing' of new members at seminars, going to professional society meetings, and writing scientific articles for publication in learned journals, will continue to be encouraged within the laboratory.

Now [if] it is not possible to measure either behaviour or outputs and it is therefore not possible to 'rationally' evaluate the work of the organization, what alternative is there but to carefully select workers so that you can be assured of having an able and committed set of people, and then engaging in rituals and ceremonies which serve the purpose of rewarding those who display the underlying attitudes and values which are likely to lead to organizational success, thus reminding everyone of what they are supposed to be trying to achieve, even if they can't tell whether or not they are achieving it?

Whereas output and behaviour control can be implemented through a market or a bureaucracy, ceremonial forms of control can be implemented through a clan. Because ceremonial forms of control explicitly are unable to exercise monitoring and evaluation of anything but attitudes, values, and beliefs, and because attitudes, values, and beliefs are typically acquired more slowly than are manual or cognitive abilities, ceremonial forms of control require the stability of membership which characterizes the clan.

Thus, in the context of the strategic implementation of complex technical systems, we also must rely on selection and training, socialisation and a 'clan-like' culture to help ensure those involved develop and display the attitudes, values and beliefs that are likely to lead to success. What are these appropriate attitudes, values and beliefs that are likely to lead to success? Vlasic and Yetton conclude their study by discussing four capabilities for success that enable organisations to reduce project uncertainty (adapted from Vlasic and Yetton, 2002, pp. 9–10):

1 *Dynamic capabilities*: organisations must be able to adapt to meet changing environmental needs. Given a highly uncertain market organisations must be able to eliminate some of that uncertainty by creating an internal capability. When facing high uncertainty and changing environments, the variations in projects over a portfolio, and over time, create a situation where firms need to be able to adapt to situations, and to learn from them, as they arise. Where possible, this learning can be best captured and transferred, with the stability of membership that characterises the clan.

2 *Organisational competency*: firms cannot be expected to have all the skills required to successfully complete a complex change project. They need to focus on the critical elements where they add value. For example, in the construction industry it is clear that different organisations have specific roles such as architecture, engineering and subcontracting. Organisations make no attempt to have all the necessary skills to complete a large and complex project. Organisations should focus on how a particular technology adds strategic value, for example, and leave a concern for the latest developments in the technology to an IT vendor. Important issues for an organisation to focus on might be their ability to lead the project, to understand the strategic objectives of the new technology or their ability to integrate their marketing, business and IT strategies.

3 *Project management skills*: project managers are 'kings' in the construction industry. There is a clear need for the development of change project managers. People must be given the opportunity to develop those skills. Technically competent people do not necessarily make good project managers. We discussed this topic in Chapter 7.

4 *Support network for project managers*: to support project managers, a network must ensure that they have what they need. This network must also be flexible as project managers need changes over time. Organisations can create project-centred structures or develop a career structure for project managers that is independent of the technical path followed by most.

A Balanced Measurement System

Once you know what you are going to measure (i.e., the inputs, the process and/or the outputs) other important considerations are the level of analysis at which you want to measure (i.e., individual, group, organisational, societal) and whether you should focus on financial or non-financial measures. The second characteristic of effective measurement-oriented organisations, according to Lingle and Schiemann (1996), is that they balance financial and non-financial measurement.

Financial Measures: Measuring the Costs of Change

Measuring the costs and benefits of a change programme is similar to trying to measure the costs and benefits of a training programme. In 1959, Donald Kirkpatrick identified four levels of training evaluation that can be applied equally well to the evaluation of change in organisations:

1 Reaction (i.e., training: do they like it? What does the learner think and feel about the training? Organisational change: what do various stakeholders think and feel about the change inputs, process or outcome?)

2 Learning (i.e., training: do they get it? What facts, knowledge, etc., did the learner gain? Organisational change: do stakeholders think differently as a result of the change(s)? Has the change affected the way stakeholders perceive the situation? Has it taught them something?)

3 Behaviour (i.e., training: can they do it? What skills did the learner develop? What new information is the learner using on the job? Organisational change: In what way(s) has the change(s) affected people's behaviour? Do stakeholders actually do some thing(s) differently and behave differently as a result of the change(s)?)

4 Results (i.e., training: do they use it? Does it make a difference? What results occurred? Did the learner apply the new skills to the necessary tasks in the organisation and, if so, what results were achieved? Organisational change: in what way(s) does the change make a difference to organisational performance, individual performance, costs, etc.?)

Measuring change return on investment (ROI)

To determine return on investment (ROI) of a change or training programme you must measure it at Kirkpatrick's results level. To achieve results measurable in ROI terms, you must identify both the results you want and the way you will obtain data to determine costs and benefits. To determine the desired results, we need to talk to the people to be trained and their peers, superiors and subordinates. To track results, we must choose indicators that are credible. This is critical, as people must agree that the indicators you track will 'prove' (or at least provide a credible measure of) change/training ROI.

Step one   Determine the costs of the intervention. In many circumstances this has to be estimated as some costs are obvious and direct, while others are less obvious and indirect. For example, technology purchased, time spent by team members on the project, legal costs and marketing expenses are all relatively easily traced and costs can be attached. The time spent by senior management sponsors, the cost of lost productivity due to not having team members doing their normal job, and the appropriate overhead costs are much more difficult to measure accurately. In this case, a reasonable estimation (or guess) is better than nothing.

Step two   Determine the benefits of intervention. In most circumstances, this has to be estimated as the benefits of many change interventions are not easily attributable to direct revenue increases or cost decreases. Once again, a reasonable estimate (or guess) is better than nothing. Consider the following hypothetical examples:

1 Reduced accounts receivable. Because of process changes and the installation of electronic data interchange, instead of taking two to three days to generate and then post an invoice, electronic invoices can be generated within minutes of an order being placed. This will reduce the time-to-payment by some proportion that can be multiplied by the total debtor days for an estimate of money saved in a year.

2 Reduced number of invoices. If you have to send, on average, 2.5 invoices before getting paid, at a cost of at least £0.50 per invoice, the cost of sending an electronic invoice can be estimated and the cost savings, or increases, can be estimated.

3 Reduced staff turnover and absenteeism. Over the years, I have seen studies that estimate the costs of replacing a full-time, white-collar employee ranging from one-half to four times their yearly salary (cf. Richards, 2000). If you can say that a certain intervention is likely to reduce staff turnover by even 10 per cent, the cost can be estimated. For examples of how to cost staff turnover and absenteeism see Exhibit 8.2 and search the web and visit the following sites (http://www.mxl.com/ia/TurnoverCalculator.html; http://www.superbstaff.com/costcalc.htm for tools to help you estimate the costs of staff turnover).

Exhibit 8.2   Estimating the cost of staff turnover

I have used the following simple numbers: £50/hour rate, 40 hours/week and 50 weeks/year for total salary and benefits of £100,000/year.

Staff turnover costs = separation costs + replacement costs + training costs + productivity lost.

Separation costs

1 Exit interview = cost for salary and benefits of the interviewer and departing employee during the one-hour exit interview = £50 + £50 = £100
2 Administrative and record keeping = £50

Total separation costs = £100 + £50 = £150

Replacement costs

1 Two, one-day newspaper advertisements @ £3,000 each) = £6,000
2 Pre-employment administrative functions and record keeping = £100
3 Selection interviews = £600
4 Selection testing = £300
5 Meetings to discuss candidates = £600

Total replacement costs = £6,000 + £100 + £600 + £300 + £600 = £7,600

Training costs

1 Booklets, manuals, human resources policy documents and reports = £100
2 Induction (cost of salaries during the induction day) = £1,000
3 New employee education and training = £2,000

Total training costs = £100 + £1,000 + £2,000 = £3,100

Productivity lost

1 Productivity lost from the time and employee leaves to the time the new employee is hired = £4,000
2 Productivity lost from the time the new employee begins until s/he gets up to speed = £4,000

Total productivity lost = £4,000 + £4,000 = £8,000

Total staff turnover costs = £150 + £7,600 + £3,100 + £8,000 = £18,850

So, if an intervention is expected to reduce or increase staff turnover by 5%, you multiply the yearly total staff turnover number by £18,850 and that total by 0.05 to estimate the amount saved or lost due to changes in staff turnover.

Here are some results indicators for various types of interventions to look for and potentially estimate:

For change interventions likely to affect supervisory and management performance:
- increased output
- reduced absenteeism and tardiness
- reduced cost of new hires
- reduced turnover
- increased number of employee suggestions
- climate survey data (morale and attitudes)

For change interventions likely to affect sales:
- sales volume
- average sale size
- add-on sales
- close-to-call ratio
- ratio of new accounts to old accounts
- number of items per order

For change interventions likely to affect customer relations:
- accuracy of orders
- size of orders
- number of transactions per day
- adherence to credit procedures
- number of lost customers
- amount of repeat business
- number of referrals
- number of complaints

One of the more difficult aspects of linking an intervention to the 'bottom line' is verifying that the changes are actually linked to the operational results being measured. The key question you must answer is: 'Did the change actually cause the results, partially or completely, or were the results due to some other factors that had nothing to do with the changes?'

Sometimes it is easy to show that an intervention caused a result. If a new piece of software makes it possible to do something in ten minutes that used to take two hours manually or if a training course taught machine operators to repair a machine and the downtime figures for the following three months show less machine downtime, the causal link is relatively straightforward. Proving that a change in organisational culture or morale or an interpersonal skills training course caused an increase in sales and customer retention or reduced conflict within a sales team, however, can be more difficult.

One approach to calculating the ROI of 'soft skills' training is to observe successful performers. By observing the 'stars', and the conditions under which they work, you can establish benchmarks or job standards. Then observe average performers and the conditions under which they work. This may provide clues to the skills or behaviours that separate average from 'star' performers. Teaching the skills that create 'star' performers, and creating the same conditions, should increase results. If this happens, then you can be relatively comfortable that those 'softer' factors directly contributed to the bottom line changes.

Step three   Calculate ROI. Once you have estimates of the costs and benefits of your intervention, you can use the following formula to calculate the ROI:

$$ROI = (Benefits - Costs)/Costs \times 100$$

For example, if you determine the benefits of a training programme or change intervention are £200,000 and the costs of the programme were £50,000, the calculation would look like this:

$$ROI = (200,000 - 50,000) = 150,000/50,000 = 3 \times 100 = 300\%$$
return on investment

Non-Financial Measures: Measuring the Effects of Organisational Change

As discussed in Chapter 3, when we first looked at developing a business case and the issues of cost justification and human due diligence, we saw that balanced measurement systems are best. When considering measures of the success of strategy or new technology implementation balance is also important. The 'modern classic' in this area is the work done by Kaplan and Norton (1996) on the balanced scorecard. If you are not familiar with the concept, the basic idea is that a balanced scorecard extends traditional organisational-level financial measures, which bear little relation to progress in achieving long-term strategic objectives, to consider measures related to customers, internal business processes and learning growth.

The four balanced perspectives from Kaplan and Norton (1996) are shown below:

1 Financial: how do we look to shareholders? Measures such as operating profit, ROC (return on capital), EVA (economic value added), growth in equity, and intangibles such as the value of the brand.
2 Customer: how do we look to customers? Measures such as customer satisfaction, customer retention, customer loyalty, levels of trust, service level, and image.
3 Internal execution: what must we excel at? Measures such as new product/service development, quality, cycle-time, internal efficiency and costs.
4 Innovation and learning: How can we continue to improve and create value? Levels of education and training, new applications of information technology, employee satisfaction and retention levels.

According to Kaplan and Norton (1996), adding the customer, as well as the internal execution, innovation and learning perspectives to the traditional financial perspective provides some balance between:

● the short-term (financial) and long-term (innovation/learning)
● the external (financial and customer) and internal (execution and learning)
● the outcomes (financial and customer) and drivers (execution and learning)
● the objective, tangible (financial and execution) and subjective, intangible (customer and learning)

Kaplan and Norton (1996) suggested a critical factor is that the balanced scorecard helps companies connect long-term objectives with short-term actions by means of the following:

1 Translating the vision into operational terms that provide useful guides to action at the local level to help managers build a consensus around the organisation's vision and strategy.
2 Communicating their strategy up and down the organisation and linking it to departmental and individual objectives. Traditionally, departments are evaluated on their financial performance, and individual incentives are tied to short-term financial goals. A balanced scorecard helps managers ensure people at all levels of the organisation understand the

long-term strategy and that both departmental and individual objectives are aligned with it.

3 Business planning enables companies to integrate their business and financial plans. Almost all organisations today are implementing a bewildering array of change programmes, each competing for senior executives' time, money and other resources. Managers find it difficult to integrate those diverse initiatives and ensure they achieve their strategic goals. Using a balanced scorecard as the basis for allocating resources and setting priorities helps managers undertake and coordinate only those initiatives that move them towards their long-term strategic objectives.

4 Feedback and innovation gives companies the capacity for strategic learning. Typically, feedback and review processes in organisations are backward-looking as they focus on whether the company, its departments, or its individual employees have met their budgeted financial goals. The balanced scorecard, on the other hand, can be more forward-looking and help managers evaluate strategy in the light of recent performance enabling companies to modify strategy to reflect real-time learning.

According to Kaplan and Norton (2001), when companies align and focus their resources, their executive teams, business units, human resources, information technology and financial resources on their organisation's strategy a new culture emerges centred on the team effort to support the strategy. They have identified five key principles common to all successful balanced measurement companies (Kaplan and Norton, 2001, pp. 41–2):

1 They translate strategy into operational terms. Strategy cannot be executed if it cannot be understood. Each measure must be embedded in a chain of cause-and-effect logic that describes how intangible assets become transformed into tangible customer and financial outcomes. We will discuss this issue if linking strategic and operational measures in the next section of this chapter.

2 They align the organisation to the strategy. Organisations are traditionally designed around functional specialties such as finance, manufacturing and sales. Communicating and coordinating across functional lines is often difficult. Executives in strategy-focused organisations replace formal reporting structures with strategic themes and priorities that enable a consistent message and a consistent set of priorities to be used across diverse and dispersed organisational units.

3 They make the strategy everyone's everyday job. Strategy-focused organ-isations use balanced measures in three distinct processes to align their employees to the strategy: communication and education, developing personal and team objectives, and incentive and reward systems.

4 They realise strategy is a continual process, not a one-off event. They are able to adapt their strategies as the world changes or the strategy matures. Three themes are important in implementing such a process: (1) linking strategy and budgeting by setting stretch targets and scorecard-based strategic initiatives, often replacing fixed budgets with rolling forecasts; (2) closing the strategy loop with feedback systems linked to the score-card; and (3) testing, learning and adapting strategy based on feedback from scorecard measures.

5 They mobilise change through executive leadership. If those at the top are not energetic leaders of the process, change will not take place, strategy will not be implemented, and the opportunity for breakthrough perfor-mance will be missed. Initially, the focus is on mobilisation and creating momentum. Next, the focus shifts to governance: executives must estab-lish a process to guide the transition to a new performance model. Finally, and gradually over time, a new management system evolves.

Kaplan and Norton (2001, p. 42), suggest that this 'practical and proven framework helps solve a universal management problem – not just how to formulate strategy, but how to make it work. It can help today's leaders shape their companies to meet the challenges and reap the rewards of a new competitive era.'

By focusing solely on financial measures, we also tend to over-rely on measures at the organisational level of analysis. Carlopio (1998) suggested we consider at least four levels of analysis when trying to measure the suc-cess of implementation: (1) the individual, (2) the group, (3) the organisation, and (4) the greater community and society within which the organisation operates.

Linking Strategic and Operational Measures

The third characteristic of effective measurement-oriented organisations is that they clearly link strategic and corporate-level objectives and measures to operational and department-level objectives and measures (Epstein and Westbrook, 2001; Kaplan and Norton, 1996, 2001; Lingle and Schiemann, 1996). Many organisations routinely collect corporate-level business meas-ures related to financial reporting (e.g., return on net assets, or RONA) and key performance indicators (e.g., revenue and sales targets, staff numbers, meeting or exceeding forecasts). At the individual, group and department levels, however, it is frequently difficult to link what happens on a daily basis with these corporate-level measures. Without these clearly articulated links, people cannot easily see how what they do on a day-to-day basis affects corporate-level, strategic objectives.

In other words, while meeting individual-level and departmental goals is the focus of many employees' daily activities, it will not lead to greater corporate profit or other strategic-level objectives if the links between them are not clear. According to Epstein and Westbrook (2001), for example, at many banks, 'profit is driven by customer behavior, which is driven by employee commitment, which is influenced by leadership'. Using this conceptual model, and considerable amounts of marketing data, they suggest an organisation can define critical customer loyalty variables (e.g., intentions to continue using or purchase more services). These variables can be linked to various drivers (e.g., customers' perceptions and barriers to switching). Finally, drivers for each of the loyalty drivers can be identified (e.g., core-services loyalty is linked to customers' evaluation of local branches and efficiency in resolving problems). These types of model can be used to help managers identify key relationships between variables, and their measures, at the operational and strategic levels (e.g., a 1-point increase in loyalty-behaviour elements increases profits by £0.60 per month per customer).

Consider the following example based on some activities and local, operational-level measures that may apply to fleet management:

Activity	Local measures
(1) Vehicle acquisition, and maintenance of registration, insurance, fleet cars and repairs	Petrol costs Maintenance costs
(2) Vehicle disposal and off-lease advice	Disposal profit/loss
(3) Accident-handling and driver training	Average cost per accident, accident frequency

Now, if we link each of the local operational measures to corporate-level financial outcomes we can more clearly see how what we do as employees managing a fleet of automobiles directly affects various important outcomes (see Figure 8.2).

Now, let us take a look at an example where we try to link variables related to the implementation of innovation and technical change to existing corporate-level measures (see Figure 8.3).

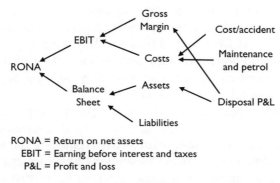

RONA = Return on net assets
EBIT = Earning before interest and taxes
P&L = Profit and loss

Figure 8.2

Figure 8.3

If we can draw links between variables such as those listed at the right side of Figure 8.3 (e.g., number of e-mail messages sent, customer retention rates) and the corporate-level financial measures listed on the left, we can clearly see what the financial effects of new technology that shortened delay-to-answer for a customer inquiry would be.

When to Measure: Feedback and Measurement

The fourth and fifth characteristics of effective measurement-oriented organisations, according to Lingle and Schiemann (1996), are that they update their strategic 'scorecard' regularly so that information and measures stay current, and they clearly communicate measures and progress to all employees. As mentioned previously in this chapter, feedback is an essential process allowing people to test strategy and technical innovation, learn from their experiences and adapt (Kaplan and Norton, 1996, 2001; Legge, 1984).

Within the fields of cybernetics and general systems theory, feedback equals control. Feedback is the return of part of a system's output to affect its input. Consider a thermostat trying to control the temperature in a room, for example. The thermostat senses the room temperature (feedback from the environment) and adjusts the function of the heating/cooling units to change the room's temperature until it senses the room has achieved the desired temperature. This feedback process is what provides control in any system, whether it is a thermostat sensing the room temperature or people in an organisation getting information from its environment (i.e., employees, customers, competitors, the stock market). This feedback process allows managers to exert some level of control over the process. Without feedback in a system, there is no control, no learning and no change. (For more information on cybernetics and general systems theory, see the following readings and websites: Ashby (1956); S. Beer (1981); Bertalanffy (1962); http://artsandscience.concordia.ca/edtech/ETEC606/feedback.html; http://pespmc1.vub.ac.be/DEFAULT.html.)

Just as a thermostat must regularly measure the room's current temperature, compare it to the desired temperate and make adjustments, managers must regularly measure a range of variables (across the individual, group, organisation, and societal levels from the financial, customer, internal business processes, and learning/growth perspectives), compare the results to the desired levels, and make adjustments. A question that naturally follows this line of thinking is: how 'regular' is 'regularly'? The best answer I can give is not really satisfactory: feedback should be frequent. How 'frequent', is 'frequent', you might justifiably ask? With feedback, as with many good things in life, my general suggestion is to try to get it as often as you reasonably can. I can say for sure that once a year is not enough, but three times a day is probably over-doing it. You will have to use your judgement and consider the costs, the time involved and other factors when answering this question in reality.

Beyond Balance

It is important to remember that a measurement system operates within a larger organisational context. Research suggests there are four mechanisms that contribute to the success of measurement-managed organisations (cf. Lingle and Schiemann, 1996).

First, measurement-managed organisations have agreement on strategy. Only 7 per cent of the measurement-managed companies in Lingle and Schiemann's study reported a lack of agreement among top management on the business strategy of the organisation compared to 63 per cent of the non-measurement-managed organisations. The act of translating vision or strategy into measurable objectives forces specificity. It helps to bring to the surface and resolve those hidden disagreements that often get buried when the strategy remains abstract, only to return at some later date to haunt managers.

Second, measurement-managed organisations have clarity of communication. Lingle and Schiemann asked managers how well their business strategy was communicated and understood from top to bottom in their organisations. Sixty per cent of managers in measurement-managed organisations thought the strategy was well communicated and understood, while only 8 per cent of managers from non-measurement-managed organisations thought the strategy was well communicated and understood.

Third, measurement-managed organisations ensure there are explicit links between corporate-level, strategic measures, and more local, operational-level measures. Measurement-managed organisations reported more frequently that unit performance measures were linked to strategic company measures (74 per cent versus 16 per cent for non-measurement-managed companies) and that individual performance measures were linked to unit measures (52 per cent versus 11 per cent, respectively).

Finally, measurement-managed organisations have organisational cultures that help provide a set of collective attitudes and behaviours that help sustain competitiveness. Take teamwork, for example. When compared to their non-measurement-managed counterparts, managers of measurement-managed companies more frequently reported strong teamwork and cooperation among the management team (85 per cent favourable ratings on teamwork versus 38 per cent for non-measurement-managed companies). Lingle and Schiemann also asked respondents the extent to which employees in their organisation self monitored their performance against agreed-upon standards. Forty-two per cent of measurement-managed companies reported excelling in this regard compared with only 16 per cent of their non-measurement-managed counterparts. Finally, 52 per cent of managers from measurement-managed companies said employees generally were unafraid to take risks to accomplish their objectives, compared with only 22 per cent of the non-measurement-managed companies.

CHAPTER 9

Beyond Implementation

Finally we have come to the end, only to realise that it is a new beginning. The issues of project termination, continuous improvement, innovation and the learning organisation must all be considered at this point.

In this chapter we begin by juxtaposing the issues of project termination and continuous improvement and asking the question, 'Do implementation projects really ever end?' We then compare the benefits of continuous improvement efforts and radical innovation and conclude that we need them both. Within this conversation we look at several ways to stimulate both continuous improvement and radical innovation (i.e., via trial and error, innovation hubs and clusters, and the effects of various elements of organisational culture). Finally, we take a look beyond the strategic implementation of technology at how chaos theory and complexity theory can be applied to organisations.

Project Termination or Continuous Improvement?
Cutting your Losses

As discussed in Chapter 6, commitment is especially important when trying to innovate. While we usually are, therefore, concerned with how we can increase levels of commitment to change, over-commitment to a course of action can also become a problem. Day and Schoemaker (2000) illustrated that many organisations are reluctant to give up (that is, they over-commit to) an existing, known technology, are reluctant to commit fully to a new technology and lack persistence when trying to exploit emerging technical opportunities. Their research highlights that commitment has both a positive and a negative side. In other words, while not enough commitment is certainly a problem in many situations, over-commitment can be a problem as well.

If you are human, you are subject to the phenomenon of 'escalation of commitment to a failing course of action' (Keil and Montealegre, 2000). Especially when involved with technologically sophisticated projects, instead of terminating or redirecting the failing endeavour, managers frequently continue pouring in more resources. Information technology projects with high complexity, risk and uncertainty are particularly susceptible to escalation. The solution is to recognise the problem, objectively re-examine the previously chosen path, search for an alternative course of action and implement an exit strategy. While this advice may seem simplistic, there is more to it. The first step is the hardest. You know you have a serious problem when you notice mounting external pressure that can no longer be ignored and you start to think of the project as an 'albatross', a 'bottomless pit' and/or a 'millstone around your neck'. Because it is sometimes impossible for those involved to be objective, you might need to engage an external agent to obtain independent evidence of problems with the existing course of action and to clarify its magnitude. Although you might have made up your mind by this point, you cannot recommend cancelling the project without offering an acceptable alternative. Finally, key stakeholders must be prepared and impressions must be managed. The opportunity costs and ambiguity involved in continuing should be contrasted with the certainty gained by cancelling the project.

There are several 'classic' reasons why projects may be formally terminated (Roman, 1986):

- *Convenience*

Termination for convenience is a difficult decision. There is often the connotation of project failure. Sometimes business objectives change, senior executives or management personnel change, markets and technology change, cost expectations are over-run, progress is too slow and so on, and the decision must be taken to discontinue the project. If this happens, it is best to manage the process right away by communicating the truth to everyone involved and affected. There is always the potential for serious negative reaction when a project is terminated for convenience. People frequently try to lay blame, to clear their names and reputations from the failure, and to make it seem that it was not a decision actually made by choice but one that the organisation was forced into. It is important to remember that it is rare that a failed project is a total waste of time, money and effort. People at the very least gain experience, as well as the technical and non-technical knowledge that is generated and, if managed properly, can be captured and learned from.

- *Default*

Projects may also be terminated by default. Unsatisfactory technical performance, contract and legal violations, delivery delinquencies and quality discrepancies are all valid reasons for project termination by default. In this case, the project has gone beyond the boundaries and scope of what is considered as acceptable as is terminated.

- *Completion*

If the implementation was a success, the point is eventually reached where the implementer needs to hand over the project to those responsible for 'normal' operations. For a consultant, this may mean handing the project back to the organisation and terminating involvement with the project. Project completion is a process, not a single event. This is where the roll-out process blends into the confirmation/measurement and continuous innovation/improvement stages.

Of course, all projects that begin must have an end. This may or may not be the case. Let us consider the following paragraph from a chapter in Carlopio (1998):

> We now enter a never-ending cycle of implementation, evaluation, modification and evolution. Many people have thought that although the process of implementation may go on for a considerable period of time, depending on the circumstances and the technology, it must end eventually. At some point, it was thought, the innovation or new technology becomes institutionalised and viewed as part of stable operations. Many workplace innovations and new technologies, however, seem to change all of this. It is best now to consider that the process of implementation never actually ends, as some people are reinventing continually and looking for ways of improving their processes incrementally.

In this view, as part of the roll-out efforts, plans must be made for the evolution of the innovation or new technology. Progress is monitored, tests are conducted, breaking-in and de-bugging are attended to, and any necessary modifications are made to the innovation or new technology. The roll-out phase blends seamlessly into confirmation and measurement activities, which in turn blend into this final phase of adjustment, innovation and continuous learning and improvement. The process of re-invention, discussed in relation to the topic of an innovation analysis (see Chapter 5), becomes especially relevant at this point. Re-invention was defined as the degree to which an innovation is changeable or modifiable by the user(s). In the process of their adoption and implementation, certain innovations will be more or less modified by the users. Issues such as customisation and incremental improvements via experimentation and adjustments, therefore, are relevant at this point as well.

Continuous Improvement or Radical Innovation?

Do we need processes that allow us to evolve and continuously improve systems and technology or do we need radical innovation and significant improvements? A colleague of mine is fond of saying, 'When given the choice between two things, take both.' I agree, and so do several others.

Mathews, White and Long (1999) state that change may be either incremental and evolutionary or discontinuous and revolutionary. Unfortunately, they continue, researchers often tend to emphasise one or other mode of change (e.g., evolutionary versus revolutionary) as the sole explanation of the observed dynamics in a particular situation. This, according to Mathews, White and Long, results in an 'either/or' fallacy. In other words, change is sometimes evolutionary and sometimes revolutionary. When we look at a situation and try to explain it, it sometimes looks like gradual change and other times it looks like sudden change. We tend to think, therefore, that change is either gradual or sudden, when in fact change is sometimes incremental and sometimes radical.

Hamel (2001) suggests continuous improvement and radical innovation are in no way mutually exclusive. Radical innovation and incrementalism must go hand-in-hand because each is valuable in its own right. Hamel (2001, p. 150) states, 'The quest to get better, day by day, can hardly be faulted. I would hope that every employee in every company strives constantly toward this goal.' Radical innovation, on the other hand, is more likely to create large amounts of new wealth. Hamel (2001) also reminds us that radical innovation does not imply jettisoning the past. While the central challenge of radical innovation is to learn how to escape the orthodoxies and dogmas that blind us to new opportunities, at the same time we must leverage the brands, assets and competencies we have built up over time, not abandon them. Another mistake Hamel points out is thinking that radical innovation is more risky than continuous improvement. Incrementalism is not low-risk, as 'the status quo' often represents the biggest risk of all. 'While incrementalism may reduce the company's immediate investment risks, it often raises its long-term strategic risks. Conversely, radical innovation need not be high risk' (Hamel, 2001, p. 151). While radical innovation does lead into areas where people have much to learn and their old assumptions need to be challenged, this need not imply high risk. What it does imply is the need to adapt and to learn quickly. This can be achieved, according to Hamel, through a series of low-cost, low-risk experiments and market incursions.

The logical question to ask at this point is, 'How can we stimulate continuous improvement, radical innovation and learning in our organisations?' This question is especially important if we work in large, mechanistic organisations. Research indicates that improvement, innovation and learning can be stimulated in a number of ways.

Trial and Error: Innovation via Experimentation

Consider the following from Thomke (2001, pp. 67–8), who seems to agree with Hamel about the importance of experimentation in successful innovation:

> Experimentation lies at the heart of every company's ability to innovate ... today, a major development project can require literally thousands of experiments, all with the same objective: to learn whether the product concept or proposed technical solution holds promise for addressing the new need or problem, then incorporating that information in the next round of tests so that the best product ultimately results ... new technologies such as computer simulation, rapid prototyping, and combinatorial or chemistry allow companies to create more learning more rapidly, and that knowledge, in turn, can be incorporated into more experiments at less expense.

Brown and Eisenhardt (1997) found that successful managers improvise and explore potential innovations by experimenting with a wide variety of low-cost probes. Their results suggest these multiple product innovation practices form a core capability central to success. Kanter (2001) suggests it is better to launch many small experiments and learn from the results of each than to make one big bet. Beinhocker (1999) suggests organisations should not have one focused strategy, but should cultivate and manage multiple strategies that evolve over time; in other words, innovate and experiment. Hargadon and Sutton (2000) suggest that refining and putting concepts to the test via prototypes, experiments, simulations, models and pilot programmes is an essential step in the innovation process. Similarly, Sutton (2001) also lists experimentation as a key to successful innovation. He suggests we need to develop organisational cultures that support constant experimentation, generating and testing many ideas, and promoting a constant and constructive contest wherein the best ideas eventually win. The traditional, mechanistic, top-down strategies for improvement, innovation and learning simply cannot keep pace today. Experimentation and rapid learning is the better approach (Pascale, 1999).

Thomke (2001) has identified the essentials for what he calls enlightened experimentation. He suggests that successful innovation requires the right organisational systems for performing experiments that can generate the information needed to develop and refine innovations quickly. The challenges are both managerial and technical, he suggests. Thomke's four essentials are (2001, p. 69) listed below:

1. Organise for rapid experimentation. Ensure that organizational routines, boundaries, and incentives all encourage rapid experimentation. Run experiments in parallel instead of sequentially. Consider using small development groups that contain people with all knowledge required to iterate rapidly.

2. Fail early and often, but avoid mistakes. Do not punish failures, as experimentation, by definition, is a process of trial and error. Remember failures that occur early in the development process can significantly advance your knowledge of the keys to later success. Practice the basics of good experimentation. Design tests well, have clear objectives, and hypotheses. Be sure to control variables that could confound the results of your experiments.

3. Anticipate and exploit early information. In general, the earlier a problem is identified and solved in the development process the less expensive it is to fix. If a project fails after years of effort and huge expense, it can be devastating. Early in the process, therefore, it may be best to run a larger number of less expensive, 'quick and dirty' experiments. Complex, elegant and expensive experiments are likely best suited later in the process.

4. Combine new and traditional technologies. The true potential of many new technologies can be exploited only after an organisation can reconfigure its processes and systems to use them in concert with traditional technologies (Thomke, 2001).

Hubs and Clusters: Innovate via Collaboration

A number of recent studies suggest that creating hubs, clusters or communities of creation, both internally and externally, will stimulate innovation and learning. Stringer (2000) suggests creating internal idea markets. Most large organisations are not radical innovators as they are designed to preserve the status quo. Their structures, cultures and people do not seem to learn fast enough and well enough to successfully commercialise radical new ideas. Industry leaders cannot afford to embrace radical innovation as it is too expensive. Large organisations with bureaucratic structures and mechanistic cultures discourage bringing big ideas to market, they rely too much on internal research and development (R&D), and they do not attract or retain radical innovators (Stringer, 2000). Innovation in larger companies can be stimulated internally by making breakthrough innovation a strategic and cultural priority and by hiring more creative and innovative people.

Large organisations can grow flexible, informal project laboratories as well as create and protect 'idea markets' within the traditional organisation, and thus become more 'ambidextrous': that is, better able to simultaneously run their core business and promote radical innovation. Larger organisations often have the resources to experiment with acquisitions, joint ventures, cooperative ventures, and alliances with outside innovative entities. They can try to spin off innovative units and/or invest in new companies directly or via participation in an emerging industry fund.

Leifer, Colarelli O'Connor and Rice (2001) similarly suggest that radical innovation hubs and human resource strategies are key. They suggest that radical innovations are critical as they can restructure marketplace economics and provide the next generation of products and services. Radical innovations provide the engine for long-term growth. Leifer *et al.* (2001) conducted a six-year study of 12 radical-innovation projects in 10 large, mature companies and provided some key strategic imperatives for success. First, build a radical-innovation hub that can oversee and help nurture projects by reducing uncertainty without increasing bureaucracy: 'A radical-innovation hub can serve as a repository for cumulative learning about managing radical innovation, and is a natural home base for those who play pivotal roles in making radical innovation happen.' Second, deploy hunters and gatherers as good ideas can come from anywhere. Third, monitor and redirect projects to reduce uncertainties, do not try to over-control and continually crisis-manage. Fourth, develop a resource-acquisition skill set as radical-innovation projects typically outstrip available research resources. 'Getting money, facilities, and people is universally difficult for radical innovators; they must spend an inordinate percentage of time and energy chasing resources', and this detracts from their ability to innovate. Fifth, accelerate project transition from R&D to operational units. Finally, recruit, develop and retain people who drive radical innovation.

Not only are hubs and clusters important inside our organisations, but research suggests they are critical externally as well. The drivers of innovation, according to Harvard's strategy guru, Michael Porter, and MIT's Scott Stern, are not just internal. Although the importance of internal factors is undeniable, according to Porter and Stern (2001), the external environment for innovation is at least as important. Their research has identified three important elements of a nation's innovative capacity. The first element is a common innovation infrastructure that sets the basic conditions for innovation. For example, people with appropriate levels of education and technical skills, available financial resources, sensitive public policy (e.g., intellectual property protection, tax-based incentives for innovation, anti-trust enforcement issues, and openness of the economy to trade and investment), and the economy's level of technological sophistication are all important factors: 'A strong common innovation infrastructure requires national investments and policies stretching over decades.' The second critical element is a cluster-specific environment for innovation. Their research reveals that innovation and the commercialisation of new technologies takes place disproportionately in clusters (i.e., geographic concentrations of interconnected companies and institutions in a particular field). These clusters work because of the presence of high-quality and specialised inputs, a context that encourages investment together with intense local rivalry, the pressure and insight gained from sophisticated local demand, and the local presence of related and supporting industries. Finally, the third element is the quality of the linkages between the two elements above. These linkages are achieved by the presence of 'institutions for collaboration' such as universities that provide a bridge between R&D and technology on the one hand, and companies and commercialisation opportunities on the other.

Sawhney and Prandelli (2000) suggest that organisational communities of creation are the answer to successful innovation. Traditionally, they argue, innovation is managed internally via a hierarchical governance mechanism (e.g., an R&D department). While this closed model reduces the transaction costs that arise from coordination and retains intellectual property for the organisation, it does not allow an organisation to benefit from the creativity, diversity and agility of its external partners. Operating within a totally open governance model (e.g., Linux), on the other hand, ensures that intellectual property rights are not controlled by any single entity and the lack of strong governance and of coordination mechanisms makes such open systems unstable. Sawhney and Prandelli (2000) propose a community of creation model that captures the best of both worlds (e.g., as Sun Microsystems is doing with its Jini technology). They suggest that it is a balancing act between the organisation's needs to innovate rapidly in order to grow and its need to leverage a community's expertise while maintaining proprietary advantages. The community model strikes a balance by creating a distributed system of innovation within a group centred on an infrastructure provided by the 'developing organisation' (e.g., Sun for Jini). It consists of only those who have signed a contract that clearly defines the ground rules for cooperation, as well as members' rights and responsibilities. The main attractor is a common interest within this 'gated community' that is not open to the general public, but still benefits from diverse, interactive collaboration.

Storck and Hill (2000) also highlight the benefits of a strategic community when implementing new technology. They suggest the purposeful development of a strategic community consisting of a large group of IT professionals working at corporate headquarters and in the business is a 'new, important organizational phenomenon, of great use in a world of increasingly dispersed human resources where firms need to wisely leverage their intellectual capital'. The benefits of this option are that information technology (IT) professionals can better manage their complex infrastructure and provide high-quality, validated solutions to issues; handle unstructured problems; and deal with the never-ending new developments in hardware and software. Their data suggest that a strategic community is better able to share knowledge and filter it into business units, while the motivation for learning and developing is greater in this community structure, which should positively impact longer-term performance.

Studies in Australia also suggested that regional clustering and a national system of innovation are important elements in successful innovation. Enright and Roberts (2001) suggest globalisation has driven a paradoxical increase in development policies based on regional clusters of firms and industries. Industry clusters exist where there is a loose geographic concentration of organisations involved in a value chain producing goods and services, and innovating. Clustered organisations leverage economic advantages and synergies from shared access to information and knowledge networks, supplier and distribution chains, markets and marketing intelligence, competencies and resources. Research suggests that informal, unplanned, face-to-face communication is critical to successful innovation. Even in this technological age, geographic concentration provides a distinct advantage:

> The geographic concentrations of firms, suppliers, and buyers found in many clusters provide short feedback loops for ideas and innovations. This is particularly important for products and services that emerge through an iterative process between producer and customer, or in industries in which suppliers or buyers are important sources of new products or services.

According to the authors, much has been learned from the experiences of industry cluster development initiatives in Australia (e.g., regional clusters such as the marine industry cluster, the Adelaide Metropolitan Industry Cluster Initiative and the Australian national wine industry). Key factors contributing to successful clustering, according to Enright and Roberts (2001), are such things as leadership, vision and a long-term commitment to capacity-building. Commitment to collaboration and the development of an ambitious industry vision are also necessary. They caution that sometimes a considerable period of learning and mentoring is necessary before firms trust each other sufficiently to accept that collaboration can increase competitiveness and create opportunities. Consistent with the literature on change management is their suggestion that achieving some early success is essential to keep the momentum of the process going.

West (2001) synthesises several conclusions from studies of the drivers of innovation. His first point is that a successful national system of innovation is not a disparate set of elements; it must be a coherent system. In order to succeed, he contends, all elements of the system (i.e., technology, institutions and organisation) must be present and structured to complement one another or little benefit may be gained. Because, over the long haul, individual participants in a competitive market cannot capture sufficient returns to justify bearing the risk of radical innovation, it is necessary for the government or a not-for-profit institution to finance research and invention. The system must mobilise substantial investment resources, and devote these to inherently risky undertakings. When entering new industries, it is often necessary to diversify risk over very long time frames. Just as individual businesses seek different funding sources, depending upon their risk profile, so must efforts to develop new industry segments. The larger and more risky the investments, the broader will the funding body have to be able to diversify its positions. West concludes that Australia's national innovation system is characterised by critical gaps. There are gaps in its ability to mobilise resources, in its system for allocating investment to innovation, and most significantly in its institutions for managing the risk of science-based innovation. To build a position as a knowledge-based economy, he suggests, Australia needs to innovate in value capture as well as value creation. West thinks Australia has the skills, organisational capabilities and financial resources necessary to do so, but may lack sufficient commitment to build an institutional system capable of turning those resources into sustained innovation, prosperity and satisfying work.

A related suggestion is to stimulate innovation through knowledge brokering. Hargadon and Sutton (2000) suggest clustering works because the best innovators take an idea from one context and apply it in a different context, and the best companies have learned to systematise that process. Hargadon and Sutton's research illustrates that organisation and attitude matter more than nurturing solitary genius. Successful innovators systematically serve as brokers, linking otherwise disconnected pools of ideas. They spot old ideas that can be used in new places, new ways and new combinations. This knowledge-brokering cycle consists of four intertwined work practices: (1) capturing good ideas, (2) keeping ideas alive through 'play' and discussion, (3) imagining new uses for old ideas, and (4) refining and putting promising concepts to the test via prototypes, experiments, simulations, models and pilot programmes. Innovation and creativity are not mysterious skills inherent in a few, special individuals: they are processes that can be managed. 'If people are given opportunities and rewards for taking good ideas – as long as no laws are broken – from all sources inside or outside the company ... and apply them in new situations ... it works.'

Godin (2000) offers another explanation of why hubs and clusters stimulate innovation: it is because innovations spread like viruses. Godin suggests that the agricultural revolution was about who could build the biggest, most efficient farms. The industrial revolution was about the race to build efficient factories. He suggests the current revolution is about ideas. The problem is that we cannot farm ideas, and neither can we manufacture them in the factory. The metaphor he settles on is that ideas spread like viruses. 'Idea merchants understand that creating the virus is the single most important part of their job.' Godin suggests that the answer is to focus on creating a product and an environment that feeds the virus. If you create a radical innovation and deliver so much 'wow' and productivity that one person tells five friends, the innovation spreads like virus. Godin (2000) suggests the keys are to make your idea easy to pass on and then give people a risk-free, cost-free way to check out the idea before they commit. A radical innovation then would market itself by creating and reinforcing 'ideaviruses'. The hard part is recognising when to shift from paying to spread it, to charging users and profiting from it.

Organisational Culture: Innovate via Trust, Creativity and Communication

In addition to experimentation, and forming clusters and hubs, another set of studies suggest that there are organisational cultures, systems and processes that can be manipulated to stimulate continuous improvement, radical innovation and learning. Creativity emerges in organisations when people increase the range of available knowledge, see old problems in new ways, and break from the past. Sutton (2001) reminds us that while innovation is important to most companies, it is not, and should not be, their primary activity. 'The more routine work of making money *right now* from tried-and-true products, services, and business models' is the appropriate focus for most people. If, however, you are more concerned with the exploration of new possibilities than with the exploitation of proven knowledge, you should try some of Sutton's unusual ideas. First, hire people who will be slow to learn the company culture and do not fit in easily. People who make co-workers feel uncomfortable and do not quickly figure out how to 'do things right' are more likely to break from the past and look at things differently. Of course this will create conflict and friction; it will also provide the opportunity for new ideas that can lead to incremental for radical change. Second, encourage people to fight among themselves as well as ignore and defy superiors and peers. This 'creative abrasion' is ultimately positive. Third, reward both success and failure. Because most of us do a poor job of judging new ideas and predicting which ones will succeed, Sutton suggests the quantity of ideas is what should be managed. The key to creativity, according to this view, is to try and to keep trying. Sutton reminds us that renowned 'geniuses' usually do not succeed at a higher rate than their peers; they simply produce more (of both success and failures). Fourth, to promote creative new ideas, ignore your current customers and the marketing and sales people who represent their views. Sutton argues that listening to your current customers leads to 'more of the same' rather than to innovation and change. Instead, he suggests, we should develop a culture that supports constant experimentation, generates and tests many disparate ideas, and promotes a constant and constructive contest wherein the best ideas eventually win.

In an article that is especially relevant for organisations operating in countries overly reliant on primary industry and manufacturing, Ruppel and Harrington (2000) argue that, globally, resource-generated wealth is being supplanted by innovation-generated wealth. 'In the private sector, particularly in the high technology segment, half of today's revenues flow from products and services which are less than five years old.' The authors therefore studied the relationships between communication, ethical work climate and trust, and their ability to generate commitment to the organisation and innovation. The study illustrated that the more open communication among managers and employees was, the greater the level of trust; and the more a corporate climate emphasises human relations and employee interests, the greater the openness of employee communications. Such a result supports the idea that organisational climates that are more principle-oriented, or which have mechanisms for integration and coordination across subcultures, promote communication. Finally, their study showed that the greater the level of trust in the organisational subunit (e.g., a department), the greater was the commitment and innovativeness of the people in the subunit. This finding supports the idea that increasing trust will heighten employees' willingness to take risks and will lead to greater creativity and innovativeness.

Less monitoring and defensive behavior by managers and more employee enthusiasm for innovation are believed to be the underlying mechanisms by which trust influences innovation. Thus, this study suggests that the establishment of trust within the organization is a worthwhile effort in organizations where innovation is desired. Moreover, such effort benefits not only primary stakeholders (e.g., stockholders), but also extended stake-holders (e.g., employees).

Beyond the Strategic Implementation of Technology

Recall our discussion from Chapter 3 about whether strategy was designed or emergent. I suggested that while strategy is both emergent and designed, the rationalist, design school of thought was the implicit model used in most of the strategic planning literature and this was the view we explored in Chapters 1 and 2. In this final section, we will explore the evolutionist, emergent school of thought. In this emergent strategy view, it is suggested that instead of following a rational, linear strategy formulation and then the implementation process, managers should facilitate the involvement of many people, experiment with many options, encourage random, chaotic connections between people and ideas, and realise that the strategy formulation and implementation processes are complex and iterative. I think the emergent view of strategy will become the predominant view in the future. It certainly has different implications for the strategic implementation of technology.

A current criticism of this emergent view of strategy is that when managers try to apply it, they can end up tangled in confusion, generating multiple contingencies and perspectives that inhibit implementation and lead to failure. I think that recent work in the areas of chaos theory and complexity theory (sometimes referred to as complex adaptive systems, or CAS) is helping to solve that problem. Scientists have been struggling to understand and work with complex interactions and emergent properties without a good paradigm to guide them. Chaos and complexity theory provide a way of understanding and working with emergent properties. 'Science, it seems, has finally cracked the next level of analysis, one that will replace the Cartesian approach and substitute a new, scientific holism for the old reductionism' (Pigliucci, 2000, p. 62).

In order to understand how the concepts of chaos and complexity can help us, we must understand some of the basic chaos and CAS concepts, as well as the differences between linear, rational systems and non-linear, non-rational systems. While it is sometimes easier to understand and use mechanistic, linear, rational systems/models, much recent work suggests that organisations are actually better viewed as non-linear, non-rational, chaotic, complex adaptive systems (Bonabeau and Meyer, 2001; Carlopio, 1994, 1996; Eisenhardt and Sull, 2001; Gadiesh and Gilbert, 2001; Kurtyka, 1999; Mathews, White and Long, 1999; Pascale, 1999; Wilson, 1999).

A non-linear system is circular (e.g., it involves feedback loops) and 'open' (i.e., it interacts with its environment). It is a system in which the inputs and processes (or throughputs) are affected by feedback from the outputs and the environment. A non-rational system is not irrational; it is just not rational. Consider the following analogy from Wilson (1999, p. 30):

> Got a slight headache? Take an aspirin. Got a really bad headache? Take twenty aspirin. No don't. Aspirin has a non-linear effect. Twenty isn't twenty times better and the side effects are rather worse. Got a planning problem? Take some measurements and extrapolate. Got a real planning problem? Take lots and lots and lots of measurements and extrapolate in great detail, using all the computing power you can find. No don't. It won't work, it will paralyze the business, by the time you get the results it will be too late and the results will be wrong anyway.

A non-rational system, therefore, is chaotic (i.e., it is deterministic but not predictable). Chaotic systems are simple deterministic systems wherein the outcome is determined because the boundaries and simple operating parameters or rules are set, but the system is not predictable moment to moment or step by step as the system is subject to complex interactions.

To help us clarify some of these concepts, consider the following excerpt adapted from Pigliucci (2000):

> The birth of modern science has been attributed to a variety of circumstances, events, and people, but unquestionably one of the key figures in its development was René Descartes, the French philosopher who first articulated the fundamentals of the modern scientific method of inquiry. A major tenet of Descartes' approach was the idea that complex systems can be understood by analyzing one part at a time, and then putting things back together to yield a comprehensive picture. This reductionism has been at the core of some of the most spectacular successes of the scientific endeavor, from particle physics to molecular biology. But what if some natural phenomena simply cannot be so conveniently partitioned to facilitate our comprehension? What if breaking the components apart alters their properties so much that what we learn from the separate pieces of the puzzle gives us a different and misleading idea of the system as a whole? In other words, can reductionist science study emergent properties which, by definition, are the result of complex interactions?
> Perhaps the simplest way to understand emergent properties is to consider the relation between hydrogen, oxygen, and water. Although the

combination of two atoms of hydrogen and one of oxygen yields water, the complex properties of water (e.g., the temperatures at which it undergoes state transitions to steam or ice) are not derivable from the individual properties of hydrogen and oxygen. In other words, knowing all we know about the structure and behavior of the atoms composing water, allows us to predict the structure but not the behavior of water. This means that complexity produces new properties specific to the new level of organization (in this case, molecular vs. atomic) that are due not to the sum of the parts, but to their interaction. This, it would seem, is enough to stop the Cartesian research program dead in its tracks. What will replace the Cartesian approach?

Enter Chaos and Complexity Theory

What is chaos? In the vernacular, the word is a synonym for randomness, completely non-deterministic and irregular phenomena. Typically it carries a negative connotation: a chaotic situation is one that we would like to avoid. In mathematical theory, however, chaos refers to a deterministic (i.e., non-random) phenomenon characterized by special properties that make the predictability of outcomes very difficult. In fact, a chaotic behavior is such that even though it does not happen randomly, it looks like a series of random occurrences.

Chaotic dynamics are usually, but not always the property of non-linear systems, that is of systems whose behavior can be described by sets of non-linear equations. However, the converse is not true: not all non-linear dynamics generate chaotic behaviors. Typically, a given system of equations can produce both non-chaotic and chaotic outcomes, depending on the range of values assumed by the parameters entered into the equations. In fact, in many systems one can increase the value of a key parameter and obtain a progression of outcomes from a steady equilibrium state to regular oscillations with two equilibria, to more complex cycles with multiple equilibria, to finally bringing about the chaotic condition. Since the latter can be thought of as an ensemble of an infinite number of equilibrium points (the so-called 'strange attractor'), this process is sometimes termed the 'doubling route' to chaos.

What is Complexity Theory?

Essentially, we can think of complexity theory as an attempt to study systems that satisfy two conditions: (1) they are made of many interacting parts, and (2) the interactions result in emergent properties that are not immediately reducible to a simple sum of the properties of the individual components. This has been the goal, for example, of developmental evolutionary biology throughout the 20th century, and only very recently have researchers started to gain some significant insights into the problem.

Mathews, White and Long (1999) asked the question, 'Why study the complexity sciences in the social sciences?' That is a good question for us to consider at this point as well. They suggest that the complexity sciences (e.g., chaos theory, non-linear dynamic systems theory, the theory of self-organisation) can fill in the gaps where traditional approaches to organisational change and transformation processes are limited and have proved unsatisfactory in guiding both research and practice. The complexity sciences can help us explain current areas of controversy, paradox and equivocality in our understanding of organisational change processes. They provide a perspective that leads to a better understanding of the behaviours of people in organisational systems as they are faced with increasing uncertainty, complexity and change. Concepts such as chaos, non-linearity, feedback loops and self-organisation can be applied to organisations and to the strategic implementation of technology in many ways, including the following:

1 Hewlett-Packard displays key principles of complexity by its constant quest for new opportunities, new partnerships and alliances, and by its efforts to encourage diversity by constant experimentation and learning (Wilson, 1999).

2 'Developing strategies based on narrow predictions about the future is entirely the wrong mind-set for an inherently uncertain world. Recent scientific work suggests that, in fact, our intuition about uncertainty may be understated, and that the business world is even less predictable than we think – and that our minds are even worse at forecasting than we might hope' (Beinhocker, 1999).

3 'The Body Shop uses its franchise mechanism to create self-similar entities that adapt to local conditions while holding to a corporate ethos and value set. In a complex system, strong values are a critical success factor because you have little else to guide you' (Wilson, 1999).

4 Pascale (1999) suggests the science of complexity has yielded four principles relevant to organisational strategy:
   (a) complex adaptive systems are at risk when in equilibrium; 'in equilibrium' eventually means death;

(b) complex adaptive systems exhibit self-organisation, arising from intelligence in remote clusters or the nodes within the system, and emergent complexity generated by novel patterns;

(c) complex adaptive systems tend to move towards the edge of chaos when provoked by a complex task;

(d) One cannot direct a living system, only disturb it. Complex adaptive systems are characterised by weak cause-and-effect linkages.

5 'A feature of complexity theory that runs directly counter to much economic teaching is that instead of the laws of diminishing returns, there is often a law of increasing returns as systems "lock-in" to a particular pattern due to a very small influence at the beginning. Some observers believe that a key factor in determining the early take-up, and so the long-term success, is the amount of education and support that is given in the beginning' (Wilson, 1999).

6 Mathews, White and Long (1999) state:

one of the most significant substantive implications of the complexity sciences is that dynamic, non-linear systems may exhibit surprising and counterintuitive behaviour, making prediction and control (and possibly management as it is popularly conceived) problematic ... one implication for planners is that even if the 'rules of the game' are completely known and understood at the local level, it may be impossible to predict global results and ... planning based on prediction is not merely impractical ... it is logically impossible.

7 Wilson (1999) wrote:

Competition is about winning not defeating. In a complex system, it is often not helpful to kill off your competitors. They may well be bringing you other benefits, such as stimulating the market, bringing new ideas, keeping regulators away, providing you with allies in competition with other industries, etc., etc. Successful competition is about using competitors to create advantage for yourself. When several companies are doing this, the whole system benefits from accelerated learning and change.

Simple Strategic Rules

The most important way in which concepts from the complexity sciences can be applied to the strategic implementation of technology is in relation to the concept of strategy as simple rules. According to Eisenhardt and Sull (2001) and Gadiesh and Gilbert (2001), when the business world was relatively simple, complex strategies were winners. Now that business is so complex we must simplify.

Gadiesh and Gilbert (2001) suggest the analogy of Admiral Lord Nelson's rules of engagement as a way of understanding strategy as simple rules. Lord Nelson's simple strategic principle was, 'Whatever you do, get alongside an enemy ship.' This simple rule, according to Gadiesh and Gilbert, enabled Nelson to take advantage of the Royal Navy's seamanship, training and experience. He rejected the common practice of an Admiral attempting to control a fleet through the use of flag signals (centralised decision-making and control). Instead he gave his captains a simple rule, or strategic parameter, and left them to determine exactly how to engage:

> By using a strategic principle instead of explicit signals to direct his forces, Nelson consistently defeated the French, including a great victory in the dark of night, when signals would have been useless. Nelson's rule of engagement was simple enough for every one of his officers and sailors to know by heart. And it was enduring, a valid directive that was good until the relative naval capabilities of Britain and its rivals changed. (Gadiesh and Gilbert, 2001, p. 74)

What we need in today's organisations operating in dynamic environments are a few key strategic processes and a few simple rules to guide us through the chaos and help get strategy out of the boardroom and spread throughout an organisation (Eisenhardt and Sull, 2001; Gadiesh and Gilbert, 2001). A few simple strategic rules allow organisations to seize opportunities with disciplined flexibility. Distilling the company's corporate strategy down to a strategic principle or a simple phrase, and communicating it throughout the organisation, helps people make decisions more quickly and consistently. Dell's 'Be direct', General Electric's 'Be number one or number two in every industry in which we compete, or get out', and Wal-Mart's 'Low prices, every day' are held up as quintessential examples. Gadiesh and Gilbert (2001) suggest strategic rules are critical when a company is experiencing rapid growth. During times of rapid growth it is likely that managers are forced to make decisions about important issues for which there may be no precedent. A strategic rule can help counteract the shortage of experience. Also, during times of uncertainty and rapid change managers must react immediately to sudden and unexpected developments. A strategic rule helps ensure that decisions made by managers are consistent with each other and the overall strategy. Strategic rules provide coherence and continuity during periods of turmoil.

Strategic principles do several things (Bonabeau and Meyer, 2001; Eisenhardt and Sull, 2001; Gadiesh and Gilbert, 2001).

1 They force trade-offs between competing resource demands. They help us decide which things to do out of the thousands of opportunities we have every day.

2 They test the strategic soundness of particular actions. Simple strategic rules position a company on 'the edge of chaos', providing just enough structure to allow it to capture the best opportunities.

3 They set clear boundaries within which employees must operate while allowing freedom to innovate and experiment within those constraints. In other words, they promote disciplined flexibility and the ability to adapt to a changing environment.

4 They help make an organisation robust. Even if one or more individuals fail or makes a poor decision, the group can still perform its tasks and the organisation stays on track. In other words, they enable self-organisation. Activities are neither centrally controlled nor vocally supervised, but emerge from within the system.

While it may have been 'nice' to have a strategic principle in the past, it is a necessity for success today for those who are subject to decentralisation, rapid growth, technological change and institutional turmoil. Consider the following examples of strategic rules adapted from Eisenhardt and Sull (2001), Gadiesh and Gilbert (2001) and Bonabeau and Meyer (2001):

1 Yahoo!'s managers initially focused their strategy on the branding and product innovation processes. They lived by four strategic product innovation rules: know the priority rank of each product in development, ensure that every engineer can work on every project, maintain the Yahoo! look in the user interface, and launch products quietly. As long as developers followed these rules they could change products, work at any hour, wear anything, or bring their dogs along with them to work. These rules enabled them to innovate, quickly capitalise on popular sites and move resources where they were needed.

2 Akamai Technologies (a provider of e-business infrastructure services) uses the following rules to keep its technically-oriented staff focused on customer service: the company must staff the customer service group with technical gurus, every question must be answered on the first call or e-mail, and R&D people must rotate through customer care. These rules shaped customer service at Akamai but leave room for employees to innovate with individual customers.

3 Dell computers focuses on the process of a rapid reorganisation around customer segments. A key strategic rule for this process is that a business must be split in two when revenue hits £1 billion. This ensures that the units do not get too large and they stay focused on a specific customer segment.

4 When Cisco first moved to an acquisitions-led strategy, its strategic rule was that it would acquire companies that had less than 75 employees, 75 per cent of whom were engineers. This ensured they only acquired organisations with the right size and focus.

5 At a major pharmaceutical company, strategy centres on the drug discovery process. They use several strategic rules: researchers can work on any of ten molecules (no more than four at once) specified by the senior research committee, and a research project must pass a few continuation hurdles related to progress in clinical trials. Within those boundaries, researchers can pursue whatever they like. These rules simultaneously provide focus and freedom.

6 Miramax, the innovative movie production company, has strategic rules that guide the movie-selection process: every movie must resolve around a central human condition (e.g., love, envy), a movie's main character must be appealing but deeply flawed, movies must have a clear storyline with a beginning, middle, and end, and there is a firm cap on production costs. Within these rules, there is flexibility to move quickly and creatively.

7 At the toy company Lego, the product market entry process is their strategic focus. Lego has a checklist of rules: does the proposed product have the Lego look? Will children learn while having fun? Will parents approve? Does the product maintain high quality standards? Does it stimulate creativity? If a project fails to pass one of these tests, the business team can proceed but the hurdle must eventually be cleared. This ensures Lego's products stay focused on their core market and value proposition while allowing maximal creativity.

8 Intel focuses on manufacturing capacity allocation, given the enormous costs of fabrication facilities. Intel follows a simple rule: allocate manufacturing capacity based on the product's gross margin. According to Eisenhardt and Sull, 'without this rule, the company might have continued to allocate too much capacity to its traditional core memory business rather than seizing the opportunity to dominate the nascent and highly profitable microprocessor niche' (p. 112).

9 Nortel Networks relies on two strategic timing rules for its product innovation process: project teams must always know when a product has to be delivered to the leading customer to win, and product development time must be less than 18 months. These rules keep Nortel in sync with cutting-edge customers and keep them moving quickly on new opportunities.

10 Oticon, the Danish hearing aid company, uses the following rule to determine when to stop a product in development: if a key team member chooses to leave a project for another within the company the project is killed. This ensures a project has the people it needs to succeed.

11 EBay evolved two rules for running auctions: the number of buyers and sellers must be balanced, and transactions must be as transparent as possible. The first rule equalises the power of buyers and sellers but not who can participate, while the second rule gives all participants equal access to as much information as possible.

12 The Vanguard Group, with US $565 billion in assets under management in 2001, has differentiated itself from its competitors by following one simple rule: provide unmatchable value for the investor-owner. Because Vanguard is a mutual (rather than a public) company, they discourage frequent trades and keep their overheads and advertising costs far below the industry average. They pass the savings directly to the investor-owners.

13 At Southwest Airlines in the USA, they have a simple strategic rule: meet customers' short-haul travel needs at fares competitive with the cost of automobile travel. This rule makes the process for making important and complicated decisions about things such as what cities to service, network designs, pricing and ticketing procedures relatively straightforward.

According to Eisenhardt and Sull (2001, p. 111), there are five types of strategic rules:

- how-to-rules (e.g., every question must be answered on the first call or e-mail) spell out key features of how a process is executed
- boundary rules (e.g., companies to be acquired must have no more than 75 employees, 75 per cent of whom are engineers) focus managers on which opportunities can be pursued and which are outside the pale
- priority rules (e.g., allocation is based on a product's gross margin) help managers rank the accepted opportunities
- timing rules (e.g., product development time must be less than 18 months) synchronise managers with the pace of emerging opportunities and with other parts of the organisation
- exit rules (e.g., if a key team member leaves the team, the project is killed) help managers decide when to pull out of yesterday's opportunities

What Simple Rules are Not

It is important not to confuse a mission, vision or value statement with a strategic rule. According to Gadiesh and Gilbert (2001) a mission statement is related to a company's culture, whereas a strategic rule drives a company's strategy. Mission, vision and values statements are aspirational and sometimes use emotional language or are long and complex. Strategic rules are short, simple and action oriented.

Simple strategic rules are not broad principles or marketing slogans (e.g., 'The HP way', 'Beyond Petroleum', 'We encourage flexibility and innovation', 'Our customers are our most important asset'). These statements might contribute to an organisation's overall image and culture, but they do not guide strategic action or help individuals make decisions. Simple strategic rules are not vague. Eisenhardt and Sull provide the following example of a bank using the following guideline for screening investment proposals: our investments must be currently under-valued and have potential for a long-term capital appreciation. Eisenhardt and Sull suggest you could test if a rule is too vague by asking the question, 'Could any reasonable person argue the exact opposite of the rule?' In this case, it is hard to imagine anyone trying to argue that the company should target over-valued companies with no potential for long-term capital appreciation. In this case the rule is vague and meaningless.

Another thing a simple strategic rule is not is mindless. Eisenhardt and Sull suggest you reverse-engineer your processes to determine your implicit strategic rules. If you find, for example, when you look at your recent partnership arrangements that the rule seems to be 'Always form partnerships with small, weak companies that you can control', it may be one you want to change. If you find a rule is destroying value, rather than creating it, throw the rule out.

The last thing a simple strategic rule should be is stale. Since simple strategic rules are best used in volatile markets and changing conditions; you must continually check that your rules remain relevant and are encouraging the types of behaviour you really want. Eisenhardt and Sull discuss the example of Banc One:

> Banc One's acquisitions followed a set of simple rules that were based on experience: Banc One must never pay so much that earnings are diluted, it must only buy successful banks with established management teams, it must never acquire a bank with assets greater than one-third of Banc One's, and it must allow acquired banks to run as autonomous affiliates. The rules worked well until others in the banking industry consolidated operations to lower their costs substantially. Then Banc One's loose confederation of banks was burdened with redundant operations, and it got clobbered by efficient competitors. (p. 113)

How to Develop Simple Strategic Rules

Eisenhardt and Sull (2001) suggest most organisations need between two and seven simple strategic rules. If an organisation is young, small or too undisciplined it may need more structure and, therefore, more rules than an organisation that is older, bureaucratic, and set in its ways. In this later case, the organisation may already have too many rules that need to be replaced with a few easy-to-follow directives.

There is no standard formula for developing simple strategic rules. Trial and error, together with patience, are keys to successfully developing simple strategic rules. Since we are dealing with complex adaptive systems it is sometimes impossible to tell beforehand what the effect of a certain rule, or set of rules, will be. It is necessary, therefore, to experiment and learn from your mistakes. Patience is also necessary, as you do not want to overreact. You must give new rules time to take effect.

The following is a rough outline of a two-stage process for first developing one or a few simple strategic principles which are necessary for the development of simple operation-level rules.

Stage One

1 Draft a working strategic principle. Summarise your corporate strategy. Clarify what your basic, strategic objectives are. Determine what your one or two essential competitive advantages are. For example, are you trying to minimise costs, maximise service, most efficiently use a certain resource, increase innovation, decrease variability, and/or shorten the development process? This step is essential as it determines the ultimate objective, goal or end-state you are trying to promote with your rule(s).

2 Test its endurance. Does it capture the true essence of your unique competitive value?

3 Test its communicative power. Is it concise, clear, memorable and easy to understand?

4 Test its ability to promote and guide action. Is it too broad or vague? Does it force trade-offs, test the wisdom of a decision, and/or set boundaries?

5 Communicate it consistently, simply and repeatedly. No one can follow a rule unless they know what it is, know that you expect them to follow it and will reward them for doing so.

6 Embed it. Measure it, reward it, trust it and discuss it. Without measurement, there is no feedback and you cannot 'see' the effects of the strategic rules you have in place.

7 Test its staying power. Is it stale? Keep your eye on the principle and the behaviour(s) it seems to be encouraging. Use your measurement systems, across both financial and non-financial measures at all four levels of analysis (i.e., individual, group, organisation and community/social), to monitor the effects of your rule(s) and adapt them when necessary, while resisting the urge to tinker with them too much.

Once you have your strategy down to one, or very few simple strategic principles, you can begin the process of generating some simple, operation-level strategic rules that help people to achieve it.

Stage Two

1 What is your strategic goal or objective? The process starts with the strategic principle generated above.
2 What is the real issue? You have to determine if the pressing issue for you is operational, decisional or more related to the pace or timing of some process or activity:
   (a) how-to rules are focused on solving operational problems (e.g., coordinate activities across locations);
   (b) boundary rules help set decision parameters within which anything goes (e.g., R&D, mergers and acquisitions, investment decisions);
   (c) priority rules are focused on decisions regarding the allocation of scarce resources (e.g., time, money, attention, efforts, capacity);
   (d) timing rules help us determine the rhythm and pace of critical processes or activities (e.g., how fast/slow, when to do something);
   (e) exit rules are a particularly important type of timing rule focused on when to stop or get out of a certain activity.
3 Once you have determined the types of rules you need, you try to generate a rule or set of rules that will encourage the desired outcome(s) but not limit, or limit as little as possible, the process of achieving those ends.
4 Test their endurance. Do they capture the true essence of your unique competitive value?
5 Test their communicative power. Are they concise, clear and memorable?
6 Test their ability to promote and guide action. Do they force trade-offs, test the wisdom of a decision and/or set boundaries?
7 Communicate them consistently, simply and repeatedly.
8 Embed them. Measure them, reward them, trust them, discuss them, adapt them, etc. This is not a one-off activity: this is a continual process.

REFERENCES

Abrahami, A. (1999). 'IT investment and riskless management', *Management Services*, vol. 43, no. 4, 8–13.

Abrahamson, E. (2000). 'Change without pain', *Harvard Business Review*, vol. 78, no. 4, sp. 75.

Akin, G. and Palmer, I. (2000). 'Putting metaphors to work for change in organizations', *Organizational Dynamics*, vol. 28, no. 3, 67–79.

Arvey, R. D. and Campion, J. E. (1982). 'The employment interview: A summary and review of recent research', *Personnel Psychology*, vol. 35, no. 2, 281–322.

Ashby, W. R. (1956). *An Introduction to Cybernetics*. London: Chapman & Hall.

Bahrami, H. (1992). 'The Emerging Flexible Organization: Perspectives from Silicon Valley', *California Management Review*, vol. 34, no. 4, 33–52.

Barclay, J. M. (1999). 'Employee selection: a question of structure', *Personnel Review*, vol. 28, no. 1/2, sp. 134.

Barnes, B. K. (1993). 'Intelligent risk-taking', *Executive Excellence*, vol. 10, no. 9, 11–12.

Bateman, T. S., Griffin, R. W. and Rubinstein, D. (1987). 'Social Information Processing and Group-Induced Shifts in Responses to Task Design', *Group and Organization Studies*, vol. 12, no. 1, 88–108.

Bazerman, M. H. (2002). *Judgement in Managerial Decision Making* (5th edn). New York: Wiley.

Bazerman, M. H. and Loewenstein, G. (2001). 'Taking the bias out of bean counting', *Harvard Business Review*, vol. 79, no. 1, 28.

Beer, M. and Nohria, N. (2000). 'Cracking the code of change', *Harvard Business Review*, vol. 78, no. 3, 133–41.

Beer, S. (1981). *Brain of the Firm* (2nd edn). Chichester: Wiley.

Beinhocker, E. D. (1999). 'Robust adaptive strategies', *Sloan Management Review*, vol. 40, no. 3, 95–106.

Bell, R. R. and Burnham, J. M. (1987). 'Managing Change in Manufacturing', *Production and Inventory Management*, vol. 28, no. 1, sp. 106.

Bertalanffy, Ludwig Von (1962). 'General Systems Theory – A Critical Review', *General Systems*, 7, 1–20.

Bolman, L. G. and Deal, T. E. (1997). *Reframing Organizations*. San Francisco, CA: Jossey-Bass.

Boehm, B. (1991). 'Software Risk Management: Principles and Practice', *IEEE Software*, vol. 8, no. 1, 32–41.

Bikson, T. K. and Gutek, B. A. (1983). *Advanced Office Information Technology*. Santa Monica, CA: The RAND Corporation.

Bonabeau, E. and Meyer, C. (2001). 'Swarm intelligence: A whole new way to think about business', *Harvard Business Review*, vol. 79, no. 5, 107–14.

Boulton, R. E. S., Libert, B. D. and Samek, S. M. (2000). 'A business model for the new economy', *The Journal of Business Strategy*, vol. 21, no. 4, 29–35.

Brimm, H. and Murdock, A. (1998). 'Delivering the message in challenging times: The relative effectiveness of different forms of communicating change to a dispersed and part-time workforce', vol. 9, no. 2/3, 167–79.

Brown, S. L. and Eisenhardt, K. M. (1997). 'The art of continuous change: Linking complexity theory and time-paced evolution in relentlessly shifting organizations', *Administrative Science Quarterly*, vol. 42, no. 1, 1–34.

Burns, S. (1996). *Artistry in Training*. Sydney: Woodslane.

Burns, T. and Stalker, G. M. (1961). *The Management of Innovation*. London: Tavistock.

Byrne, D. (1971). *The Attraction Paradigm*. New York: Academic Press.

Campbell, D. J. (2000). 'The proactive employee: Managing workplace initiative', *The Academy of Management Executive*, vol. 14, no. 3, 52–66.

Carlopio, J. (1994). 'Holism: A philosophy of organisational leadership for the future', *The Leadership Quarterly*, no. 3/4, 297–307.

Carlopio, J. (1996). 'Holistic organisational health: Curing the part by focusing on the whole', in A. Gutschelhofer and J. Scheff (eds), *Paradoxical Management*. Austria: Linde, pp. 1–20.

Carlopio, J. (1998). *Implementation: Making Workplace Innovation and Technical Change Happen*. Sydney: McGraw-Hill.

Carlopio, J., Andrewartha, G. and Armstrong, H. (2001). *Developing Management Skills* (2nd edn). Sydney: Prentice-Hall.

Chapman, S. (2001). 'It pays to have firm links', *The Australian*, August, p. 13.

Chase, R. B. and Dasu, S. (2001). 'Want to perfect your company's service? Use behavioral science', *Harvard Business Review*, vol. 79, no. 6, 79–84.

Christensen, L. T. (1995). 'Buffering organizational identity in the marketing culture', *Organization Studies*, vol. 16, no. 4, 651–72.

Cialdini, R. (1985). *Influence: Science and Practice*. London: Scott, Foresman.

Cialdini, R. (2001). 'Harnessing the science of persuasion', *Harvard Business Review*, vol. 79, no. 10, 72–9.

Claver, E., Llopis, J., Garcia, D. and Molina, H. (1998). 'Organizational culture for innovation and new technological behavior', *Journal of High Technology Management Research*, vol. 9, no. 1, 55–68.

Cleland, David (1999). *Project Management: Strategic Design and Implementation*. New York: McGraw-Hill.

Coetsee, L. (1999). 'From resistance to commitment', *Public Administration Quarterly*, vol. 23, no. 2, 204–22.

Collins, J. and Porras, J. (1998). *Built to Last*. New York: HarperCollins.

Conner, D. (1998). *Leading at the Edge of Chaos*. New York: Wiley.

Conner, D. (1999). 'Human due diligence', *Executive Excellence*, vol. 16, no. 10, 10ff.

Cooney, J. (1999). 'A review of *Leading at the Edge of Chaos: How to Create the Nimble Organization*', *Ivey Business Journal*, vol. 63, no. 3, 14.

Cope, B. and Kalantzis, M. (1997). *Productive Diversity: Management Lessons from Australian Companies*. Occasional paper no. 20, The Centre for Workplace Communication and Culture, Sydney: Haymarket.

Day, G. S. and Schoemaker, P. J. H. (2000). 'Avoiding the pitfalls of emerging technologies', *California Management Review*, vol. 42, no. 2, 8–33.

De Bono, E. (1985). *Six Thinking Hats*. Boston: Little, Brown.

De Geus, A. P. (1988). 'Planning as Learning', *Harvard Business Review*, vol. 66, no. 2, 70–4.

Deephouse, C., Mukhopadhyay, T., Goldenson, D. R. and Kellner, M. I. (1995–96). 'Software processes and project performance', *Journal of Management Information Systems*, vol. 12, no. 3, 187–205.

Denison, D. R. and Mishra, A. K. (1995). 'Toward a theory of organizational culture and effectiveness', *Organization Science*, vol. 6, no. 2, 204–23.

Devinney, T. (1999). 'Course notes for "Strategy in the technology oriented company"', Australian Graduate School of Management Executive Programme.

Donaldson, L. (1985). *In Defense of Organization Theory*. Cambridge: Cambridge University Press.

Durscat, V. U. and Wolff, S. B. (2001). 'Building the emotional intelligence of groups', *Harvard Business Review*, vol. 79, no. 3, 80–90.

Dwyer, J. (1999). *Communication in Business*. Sydney: Prentice-Hall.

Easterby-Smith, M. (1990). 'Creating a Learning Organisation', *Personnel Review*, vol. 19, no. 5, 24–8.

Eby, T. L., Adams, M. D., Russell, E. A. J. and Gaby, H. S. (2000). 'Perceptions of organizational readiness for change: Factor related to employees' reactions to the implementation of team-based selling', *Human Relations*, vol. 53, no. 3, 419–42.

Eisenhardt, K. M. (1985). 'Control: Organizational and economic approaches', *Management Science*, vol. 32, no. 2, 134–49.

Eisenhardt, K. and Sull, D. (2001). 'Strategy as simple rules', *Harvard Business Review*, vol. 79, no. 1, 107–16.

Enright, M. and Roberts, J. (2001). 'Regional clustering in Australia', *Australian Journal of Management*, vol. 26, Special Issue, 65–85.

Epstein, M. J. and Westbrook, R. A. (2001). 'Linking actions to profits in strategic decision making', *MIT Sloan Management Review*, vol. 42, no. 3, 39–49.

Evans, P. and Wurster, T. S. (2000). *Blown to Bits*. Boston, MA: Harvard Business School Press.

Fahey, L. (2000). 'Scenario learning', *Management Review*, vol. 89, no. 3, 29–34.

Frame, J. D. (1994). *The New Project Management*. San Francisco, CA: Jossey-Bass.

Fuchs, P. H., Mifflin, K. E., Miller, D. and Whitney, J. O. (2000). 'Strategic integration: Competing in the age of capabilities', *California Management Review*, vol. 42, no. 3, 118–47.

Fulmer, R. M., Gibbs, P. A. and Goldsmith, M. (2000). 'Developing leaders: How winning companies keep on winning', *Sloan Management Review*, vol. 42, no. 1, 49–59.

Gadiesh, O. and Gilbert, J. L. (2001). 'Transforming corner-office strategy into frontline action', *Harvard Business Review* (May), vol. 79, no. 5, 73–9.

Galbraith, J. R. and Nathanson, D. A. (1978). *Strategy Implementation: The Role of Structure and Process*. St Paul, Minn.: West.

Gardner, H. (1983). *Frames of Mind*. New York: Basic Books.

Gardner, H. (1993). *Multiple Intelligences: The Theory in Practice*. New York: Basic Books.

Glascoff, W. D. (2001). 'Beyond the hype: A taxonomy of e-health business models', *Marketing Health Services*, vol. 21, no. 1, 40.

Godin, S. (2000). Unleash your ideavirus. www.fastcompany.com, 7 August.

Goldsmith, R. E., d'Hauteville, F. and Flynn, L. R. (1998). 'Theory and measurement of consumer innovativeness: A transnational evaluation', *European Journal of Marketing*, vol. 32, no. 3/4, 340–53.

Goleman, D. (1995). *Emotional Intelligence*. New York: Bantam Books.

Goleman, D., Boyatzis, R. and McKee, A. (2001). 'Primal leadership', *Harvard Business Review*, vol. 79, no. 12, 42–51.

Gondert, S. (1993). 'The 10 biggest mistakes of SFA (and how to avoid them)', *Sales and Marketing Management*, vol. 145, no. 2, p. 52ff.

Greenfield, M. A. (1998). 'Risk management: "Risk as a resource"'. Presentation at Langley Research Center, NASA Office of Safety and Mission Assurance, http://www.hq.nasa.gov/office/codeq/risk/risk.pdf.

Hamel, G. (2001). 'Revolution vs. evolution: You need both', *Harvard Business Review*, vol. 79, no. 6, 150–6.

Han, J. K., Kim, N. and Srivastava, R. K. (1998). 'Market orientation and organizational performance: Is innovation a missing link?', *Journal of Marketing*, vol. 62, no. 4, 30–45.

Hargadon, H. and Sutton, R. I. (2000). 'Building an innovation factory', *Harvard Business Review*, vol. 78, no. 3, 157–66.

Harrington, H. J., Conner, D. R. and Horney, N. L. (2000). *Project Change Management: Applying Change Management to Improvement Projects*. New York: McGraw-Hill.

Hendersen, J. C. and Lee, S. (1992). 'Managing I/S Design Teams: A Control Theories Perspective', *Management Science*, vol. 38, no. 6, 757–77.

Higgins, J. M. (1995). 'Innovate or evaporate', *The Futurist*, vol. 29, no. 5, p. 42ff.

Hofstede, G. (1980). *Culture's Consequences*. Beverly Hills, CA: Sage.

Hofstede, G. (1981). 'Culture and organisations', *International Studies of Management and Organisation*, vol. 10, 15–41.

Hofstede, G. (1983). 'The cultural relativity of organisational practices and theories', *Journal of International Business Studies*, vol. 14, 75–89.

Hofstede, G. (1992). 'Cultural dimensions in people management', in V. Pucik, N. M. Tichy and C. Barnett (eds), *Globalizing Management*. New York: Wiley.

Hollensbe, E. C. and Guthrie, J. P. (2000). 'Group pay-for-performance plans: The role of spontaneous goal setting', *The Academy of Management Review*, vol. 25, no. 4, 864–72.

Humenick, C. (2000). 'Scenario success', *Credit Union Management*, vol. 23, no. 2, p. 14ff.

Hurley, R. F. and Hult, G. T. M. (1998). 'Innovation, market orientation, and organizational learning: An integration and empirical examination', *Journal of Marketing*, vol. 62, no. 3, 42–54.

Huy, Q. N. (1999). 'Emotional capability, emotional intelligence, and radical change', *Academy of Management Review*, vol. 24, no. 2, 325–45.

Huy, Q. N. (2001a). 'In praise of middle managers', *Harvard Business Review*, vol. 79, no. 9, 73–9.

Huy, Q. N. (2001b). 'Time, temporal capability, and planned change', *The Academy of Management Review*, vol. 26, no. 4, 601–23.

Johnson, D. J., Donohue, W. A., Atkin, C. K. and Johnson, S. (2001). 'Communication, involvement, and perceived innovativeness: Tests of a model with two contrasting innovations', *Group and Organization Management*, vol. 26, no. 1, 24–52.

Kahn, H. and Wiener, A. J. (1967). *Year 2000: Framework for Speculation on the Next Thirty-Three Years*. New York: Macmillan.

Kalakota, R. and Robinson, M. (1999). *e-Business: Roadmap for Success*. Reading, MA: Addison-Wesley.

Kanter, R. M. (1989). 'Becoming PALs: Pooling, allying, and linking across companies', *The Academy of Management Executive*, vol. 3, no. 3, 183–93.

Kanter, R. M. (2001). *Evolve! Succeeding in the Digital Culture of Tomorrow*. Boston, MA: Harvard Business School Press.

Kaplan, R. S. and Norton, D. P. (1996). 'Using the balanced scorecard as a strategic management system', *Harvard Business Review*, vol. 74, no. 1, p. 75ff.

Kaplan, R. S. and Norton D. P. (2001). 'The strategy-focused organization', *Strategy and Leadership*, vol. 29, no. 3, 41–2.

Karan, V., Kerr, D. S., Murthy, U. S. and Vinze, A. S. (1996). 'Information technology support for collaborative decision making in auditing: An experimental investigation', *Decision Support Systems*, vol. 16, no. 3, 181–94.

Keil, M. and Montealegre, R. (2000). 'Cutting your losses: Extricating your organization when a big project goes awry', *Sloan Management Review*, vol. 41, no. 3, 55–68.

Kerr, S. (1995). 'On the folly of rewarding A, while hoping for B', *The Academy of Management Executive*, vol. 9, no. 1, 7ff.

Kerr, S. (2000). 'Rewarding performance', *Executive Excellence*, vol. 17, no. 1, 4–5.

Kezsbom, D. S. and Edward, K. E. (2001). *The New Dynamic Project Management*. New York: Wiley.

Kirch, L. J. (1996). 'The management of complex tasks in organizations: Controlling the systems development process', *Organizational Science*, vol. 7, no. 1, 1–21.

Kirch, L. J. (1997). 'Portfolios of Control Modes and IS Project Management', *Information Systems Research*, vol. 8, no. 3, 215–39.

Kirch, L. J. and Cummings, L. L. (1996). 'Contextual influences on self-control of IS professionals engaged in systems development', *Accounting, Management and Information Technology*, vol. 6, no. 3, 191–219.

Kirkman, B. and Shapiro, D. (2001). 'The impact of cultural values on job satisfaction and organizational commitment in self-managing work teams', *Academy of Management Journal*, vol. 44, no. 3, 557–69.

Koehler, K. G. (1992). 'Effective change implementation', *The Management Accounting Magazine*, vol. 66, no. 5, 6.

Kotter, J. P. and Heskett, J. L. (1992). *Corporate Culture and Performance*. New York: Free Press.

Kurtyka, J. (1999). 'The science of complexity: A new way to view industry change', *Journal of Retail Banking Services*, vol. 21, no. 2, 51–8.

Laudon, K. C. and Laudon, J. P. (1996). *Management Information Systems*. Englewood Cliffs, NJ: Prentice-Hall.

Lawler, E. E. (2000). 'Pay strategy: New thinking for the new millennium', *Compensation and Benefits Review*, vol. 31, no. 1, 7–12.

Legge, K. (1984). *Evaluating Planned Organizational Change*. London: Academic Press.

Lewin, K. (1935). *Dynamic Theory of Personality*, New York: McGraw-Hill.

Leifer, R., Colarelli O'Connor, G. and Rice, M. (2001). 'Implementing radical innovation in mature firms: The role of hubs', *The Academy of Management Executive*, vol. 15. no. 3, 102–13.

Lingle, J. H. and Schiemann, W. A. (1996). 'From balanced scorecard to strategic gauges: Is measurement worth it?', *Management Review*, vol. 85, no. 3, p. 56ff.

Liu, L., Yetton, P. and Sauer, C. (2001). *Organizational Control and Project Performance: A Comparative Analysis of Construction and IT Service Companies*. Australian Graduate School of Management, University of New South Wales, Working Paper, Sydney, Australia.

Locke, E. A. and Latham, G. P. A. (1990). *Theory of Goal Setting and Task Performance*. Englewood Cliffs, NJ: Prentice-Hall.

Long, B. (2000). 'Why CMMS implementations fail', *Plant Engineering*, vol. 54, no. 6, 30–6.

Lyytinen, K., Mathiassen, L. and Ropponen, J. (1998). 'Attention shaping and software risk: A categorical analysis of four classical risk management approaches', *Information Systems Research*, vol. 9, no. 3, 233–53.

Mahadevan, B. (2000). 'Business models for Internet based E-commerce: An anatomy', *California Management Review*, vol. 42, no. 4, 55–69.

Mathews, K. M., White, M. C. and Long, R. G. (1999). 'Why study the complexity sciences in the social sciences?', *Human Relations*, vol. 52, no. 4, 439–62.

May, D. and Kettelhut, M. C. (1996). 'Managing human issues in reengineering projects', *Journal of Systems Management*, vol. 47, no. 1, 4ff.

Merrick, B. (1999). 'Make sure technology serves your business plans', *Credit Union Magazine*, vol. 65, no. 8, 84.

Miles, R. E. and Snow, C. C. (1994). *Fit, Failure, and the Hall of Fame: How Companies Succeed or Fail*. New York: Free Press.

Mintzberg, H. and Lampel, J. (1999). 'Reflecting on the strategy process', *Sloan Management Review*, vol. 40, no. 3, 21–30.

Molitor, G. T. T. (1999). 'The next 1000 years: The "Big Five" engines of economic growth', *Vital Speeches of the Day*, vol. 65, no. 22, 674–9.

Moran, Robert and Reisenberger, J. R. (1994). *The Global Challenge: Building the New Worldwide Enterprise*. New York: McGraw-Hill.

Morita, A. (1987). *Made in Japan: Akio Morita and Sony*. London: Collins.

Morris, M. H. and Trotter, J. D. (1990). 'Institutionalizing entrepreneurship in a large company: A case study at AT&T', *Industrial Marketing Management*, vol. 19, no. 2, 131–9.

Mowday, R. T., Porter, L. W., and Steers, R. M. (1982). *Employee-Organization Linkages*. New York: Academic.

Munter, Mary (2000). *Guide to Managerial Communication: Effective Business Writing and Speaking*. Upper Saddle River, NJ: Prentice Hall.

Negroponte, N. (1995). *Being Digital*. Rydalmere, NSW: Hodder and Stoughton.

Nelson, B. (1994). *1001 Ways to Reward Employees*. New York: Workman Publishing.

O'Neill, M. (1999). 'Communicating for change', *CMA Management*, vol. 73, no. 5, 22–4.

Orster, B. (1994). 'Sex role stereotypes and requisite management characteristics: An international perspective', *Women in Management Review*, vol. 9, no. 4, 11ff.

Ouchi, W. G. (1977). 'The relationship between organisational structure and organisational control', *Administrative Science Quarterly*, vol. 22, no. 1, 95–113.

Ouchi, W. G. (1979). 'A conceptual framework for the design of organizational control mechanisms', *Management Science*, vol. 25, no. 9, 833–48.

Pascale, R. T. (1999). 'Surfing the edge of chaos', *Sloan Management Review*, vol. 40, no. 3, 83–94.

Perry, T. S. (1995). 'How small firms innovate', *Research-Technology Management*, vol. 38, no. 2, 14–17.

Piderit, S. K. (2000). 'Rethinking resistance and recognizing ambivalence: A multidimensional view of attitudes toward an organizational change', *The Academy of Management Review*, vol. 25, no. 4, 783–94.

Pierce, J. L., Kostova, T. and Dirks, K. T. (2001). 'Toward a theory of psychological ownership in organizations', *The Academy of Management Review*, vol. 26, no. 2, 298–310.

Pigliucci, M. (2000). 'Chaos and complexity', *Skeptic*, vol. 8, no. 3, 62–70.

Piskurich, G. M. (1994). 'Developing self-directed learning', *Training and Development*, vol. 48, no. 3, 30ff.

Porter, M. (1998). *Competitive Strategy*. New York: Free Press.

Porter, M. (2001). 'Strategy and the Internet', *Harvard Business Review*, vol. 79, no. 3, 63–77.

Porter, M. and Stern, S. (2001). 'Innovation: Location matters', *Sloan Management Review*, vol. 42, no. 4, 28–36.

Radosevich, L. (1999). 'CIO manages change with internal communication', *InfoWorld*, vol. 21, no. 31, 76ff.

Rapp, S. J. and McCubbin, D. P. (1997). 'Implementing a purchasing card program', *TMA Journal*, vol. 17, no. 2, 30–8.

Renn, R. W., Danehower, C., Swiercz, P. M. and Icenogle, M. L. (1999). 'Further examination of the measurement properties of Leifer & McGannon's (1986) goal acceptance and goal commitment scales', *Journal of Occupational and Organizational Psychology*, vol. 72, no. 1, 107–13.

Rice, A. (2001). 'The right stuff?', www.ITrecruitermag.com, September–October.

Richards (2000). 'Firms count cost of staff turnover', UK Newsquest Regional Press, 30 October.

Riel, P. F. (1998). 'Justifying information technology projects', *Industrial Management*, vol. 40, no. 4, 22–7.

Ritov, I. (1996). 'Anchoring in simulated competitive market negotiation', *Organizational Behavior and Human Decision Processes*, vol. 67, no. 1, 16–25.

Roberts, P. (1995). 'No office and less officious—the new executive', *Australian Financial Review*, 26 September.

Robinson, G. and Dechant, K. (1997). 'Building a business case for diversity', *Academy of Management Executive*, vol.11, no. 3, 21–31.

Rogers, E. M. (1995). *Diffusion of Innovations*. New York: Free Press.

Roman, D. D. (1986). *Managing Projects*. New York: Elsevier.

Ruppel, C. P. and Harrington, S. J. (2000). 'The relationship of communication, ethical work climate, and trust to commitment and innovation', *Journal of Business Ethics*, vol. 25, no. 4, 313–28.

Sawhney, M. and Prandelli, E. (2000). 'Communities of creation: Managing distributed innovation in turbulent markets', *California Management Review*, vol. 42, no. 4, 24–54.

Schaffer, R. H. (1997). 'Beginning with results: The key to success', *The Journal for Quality and Participation*, vol. 20, no. 4, 56–62.

Schiemann, W. A. and Lingle, J. H. (1997). 'Seven greatest myths of measurement', *Management Review*, vol. 86, no. 5, 29–32.

Schmidt, R., Lyytinen, K., Keil, M. and Cule, P. (2001). 'Identifying software project risks: An international Dephi study', *Journal of Management Information Systems*, vol. 17, no. 4, 5–36.

Schmitt, N. (1976). 'Social and situational determinants of interview decisions: Implications for the employment interviews', *Personnel Psychology*, vol. 29, no. 1, 79–101.

Scott Morton, M. S. (1991). *The Corporation of the 1990s: Information Technology and Organizational Transformation*, New York: Oxford University Press.

Slater, R. (1994). *Get Better or Get Beaten!: 31 Leadership Secrets from GE's Jack Welch*. New York: Irwin.

Smith, M. (1998). 'The development of an innovation culture', *Management Accounting*, vol. 76, no. 2, 22–4.

Sosik, J. J. and Megerian, L. E. (1999). 'Understanding leader emotional intelligence and performance: The role of self-other agreement on transformational leadership perceptions', *Group and Organization Management*, vol. 24, no. 3, 367–90.

Stajkovic, A. J. and Luthans, F. (2001). 'Differential effects of incentive motivators on work performance', *Academy of Management Journal*, vol. 44, no. 3, 580–90.

Stevens, T. (2000). 'Picking the winners', *Industry Week*, vol. 249, no. 5, 27–30.

Stewart, T. A. (1994). 'Rate your readiness to change', *Fortune*, vol. 129, no. 3, p. 106ff.

Storck, J. and Hill, P. A. (2000). 'Knowledge diffusion through "strategic communities"', *Sloan Management Review*, vol. 41, no. 2, p. 63ff.

Stringer, R. (2000). 'How to manage radical innovation', *California Management Review*, vol. 40, no. 4, 70–88.

Sutton, R. (2001). 'The weird rules of creativity', *Harvard Business Review*, vol. 79, no. 9, 94–103.

Szmigin, I. and Carrigan, M. (2001). 'Leisure and tourism services and the older innovator', *The Service Industries Journal*, vol. 21, no. 3, 113–29.

Tapscott, D. (2001). 'Rethinking strategy in a networked world', *Strategy + business*, no. 24, 34–41.

Thomke, S. (2001). 'Enlightened experimentation: The new imperative for innovation', *Harvard Business review*, vol. 79, no. 2, 67–75.

Tornatzky, L. and Klein, K. (1981). *Innovation Characteristics and Innovation Adoption-Implementation: A Meta-Analysis of Findings*, Washington, DC. National Science Foundation, Division of Industrial Science and Technological Innovation.

Trompenaars, F. and Hampden-Turner, C. (1998). *Riding the Waves of Culture: Understanding Cultural Diversity in Global Business*. New York: McGraw-Hill.

Tucker, M. L., Sojka, J. Z., Barone, F. J. and McCarthy, A. M. (2000). 'Training tomorrow's leaders: Enhancing the emotional intelligence of business graduates', *Journal of Education for Business*, vol. 75, no. 6, 331–7.

Turban, E., Lee, J., King, D. and Chung, H. M. (2000). *Electronic Commerce*. Englewood Cliffs, NJ: Prentice-Hall.

Umiker, W. (1999). 'Organizational culture: The role of management and supervisors', *The Health Care Manager*, vol. 17, no. 4, 22–7.

van der Heijden, K. (1996). *Scenarios: The Art of Strategic Conversation*. New York: Wiley.

Vecchio, R., Hearn, G. and Southey, G. (1996). *Organisational Behaviour* (2nd edn). Sydney: Harcourt Brace.

Vlasic, A. and Yetton, P. (2002). *Delivering Successful Information Technology Projects: Learning from the Experiences of the Construction Industry*. Working paper, Australian Graduate School of Management.

Walker, L. E. (2001). 'Advantages of risk-based project management', *Occupational Health and Safety*, vol. 70, no. 9, 161–3.

Weick, K. E. (1979). *The Social Psychology of Organizing*. New York: McGraw-Hill.

Weir, M. (1977). 'Are computer systems and humanised work compatible?', in Richard Ottaway (ed.), *Humanising the Workplace*. London: Croom.

West, J. (2001). 'The mystery of innovation: Aligning the triangle of technology, institutions and organization', *Australian Journal of Management*, vol. 26, 21–43.

Wetzel, J. N., O'Toole, D. and Peterson, S. (1999). 'Factors affecting student retention probabilities: A case study', *Journal of Economics and Finance*, vol. 23, no. 1, 45–55.

Whyte, G. and Sebenius, J. K. (1997). 'The effect of multiple anchors on anchoring in individual and group judgment', *Organizational Behavior and Human Decision Processes*, vol. 69, no. 1, 75–85.

Wilson, J. (1999). 'Winning through chaos – part 1', *Credit Control*, vol. 20, no. 4, 27–32.

Wood, N. (1998). 'Change champion', *Incentive*, vol. 172, no. 9, 92–3.

Yetton, P. Johnston, K. and Craig, J. (1994). 'Computer-aided architects: A case study of IT and strategic change', *Sloan Management Review*, 35, 4, 57–67.

Zajonc, R. B. (2001). 'Mere exposure: A gateway to the subliminal', *Current Directions in Psychological Science*, vol. 10, no. 6, 224–8.

Zaltman, G., Duncan, R. and Holbek, J. (1973). *Innovations and Organizations*. New York: Wiley.